Petroleum Engineering

Editor-in-Chief

Gbenga Oluyemi, Robert Gordon University, Aberdeen, Aberdeenshire, UK

Series Editors

Amirmasoud Kalantari-Dahaghi, Department of Petroleum Engineering, West Virginia University, Morgantown, WV, USA

Alireza Shahkarami, Department of Engineering, Saint Francis University, Loretto, PA, USA

Martin Fernø, Department of Physics and Technology, University of Bergen, Bergen, Norway

The Springer series in Petroleum Engineering promotes and expedites the dissemination of new research results and tutorial views in the field of exploration and production. The series contains monographs, lecture notes, and edited volumes. The subject focus is on upstream petroleum engineering, and coverage extends to all theoretical and applied aspects of the field. Material on traditional drilling and more modern methods such as fracking is of interest, as are topics including but not limited to:

- Exploration
- Formation evaluation (well logging)
- Drilling
- Economics
- Reservoir simulation
- Reservoir engineering
- Well engineering
- Artificial lift systems
- Facilities engineering

Contributions to the series can be made by submitting a proposal to the responsible publisher, Anthony Doyle at anthony.doyle@springer.com or the Academic Series Editor, Dr. Gbenga Oluyemi g.f.oluyemi@rgu.ac.uk.

More information about this series at http://www.springer.com/series/15095

Amr Mohamed Badawy ·
Tarek Al Arbi Omar Ganat

Rock Properties
and Reservoir Engineering:
A Practical View

 Springer

Amr Mohamed Badawy
Ciro, Egypt

Tarek Al Arbi Omar Ganat ⓘ
Department of Petroleum and Chemical
Engineering
Sultan Qaboos University
Al Khoud, Sultanate of Oman, Oman

ISSN 2366-2646   ISSN 2366-2654 (electronic)
Petroleum Engineering
ISBN 978-3-030-87464-3   ISBN 978-3-030-87462-9 (eBook)
https://doi.org/10.1007/978-3-030-87462-9

This Springer imprint is published by the registered company Springer Nature Switzerland AG
The registered company address is: Gewerbestrasse 11, 6330 Cham, Switzerland

I would like to dedicate this book to my family members. No words can express my gratitude for them, for they have always supported and encouraged me.

—Amr Mohamed Badawy

I would like to dedicate this book to my parents and my brothers and sisters, and to my wife Basma and my children Mohamed, Heba, Abdulrahman, and my young hero Abdul Malik. Without their support and encouragement, this book will not complete.

—Tarek Al Arbi Omar Ganat

Preface

This book was written by the authors after long overseas work experience in oil and gas industry, to handle, analyze and usage of reservoir rock properties in Field Development Plan (FDP) work. That is why the title 'Reservoir Engineering Practical View' was chosen for this manuscript. Based on our work experience as reservoir engineers in multi-regions located at different continentals, we have got much of knowledge and skills which encouraged us to share with others to make their evaluation process more easier.

Considering the rock properties which are of more interest to reservoir engineering work, this book is a trial to prepare a detailed reference to cover the subject. It is not the objective of this book to duplicate or repeat the classical reservoir rock properties courses and/or references. In the contrary, it concentrates on the importance of the reservoir rock properties in field development plan works (FDP) work, how the reservoir simulator interprets them and what the user (reservoir engineer) should do to properly introduce them to the simulator. It could be a good idea to describe this text as (*Reservoir Rock Properties; What, Why and How?*).

We hope that everybody will enjoy going through this book and find it helpful in his/her reservoir development work. It is to be mentioned that any technical matter in this book represents our best understanding. However as technical individuals, we should always consider other points of views.

Ciro, Egypt   Amr Mohamed Badawy
Al Khoud, Sultanate of Oman, Oman   Tarek Al Arbi Omar Ganat

Overview of Chapters

This book provides several technical information presented in nine chapters. All the chapters defining their content and giving the reader complete view of the rock properties which help the petroleum engineers to understand the evaluation and assessment process and the procedures required during the evaluation of hydrocarbons formation. The chapters were written in a simple way to serve as a rapid reference and guidance based on the discovery type and the amount of reservoir rock data available.

In this book, solution for the exercises introduced in different sections were presented to exemplify the stages covered in this book, permitting the petroleum engineers to understand the evaluation of the whole steps.

Contents

1   **Introduction** 1
    1.1   Trap, Reservoir, Aquifer and Basin 1
    1.2   Modeling and Simulation 3
    1.3   Reservoir Simulation Modeling 6
    References 7

2   **Coring and Core Analysis** 9
    2.1   Coring, Definition and Applications 9
    2.2   Types of Coring 10
          2.2.1   Conventional Coring 10
          2.2.2   Sidewall Coring 11
          2.2.3   Special Coring Techniques 12
    2.3   Coring Fluids 13
    2.4   Precautions While Coring, Core Handling and Core
          Transportation 13
    2.5   Routine and Special Core Analysis 14

3   **Porosity** 17
    3.1   Porosity and Porosity Types 17
    3.2   Factors Affecting Porosity 19
    3.3   Porosity Determination 21
          3.3.1   Porosity Determination, Core Measurement 22
          3.3.2   Porosity Determination, Open Hole Logs 23
    3.4   Porosity Averaging 25
    3.5   Porosity for Reservoir Simulator 27
    Reference 28

4   **Fluid Saturation** 29
    4.1   Fluid Saturation, Importance and Types 29
    4.2   Fluid Saturation Determination 30
    4.3   Fluid Saturation Averaging 31
    4.4   Fluid Saturation for Reservoir Simulator 31

5 Permeability 35
  5.1 Permeability, Darcy's Law 35
  5.2 Absolute, Efective and Relative Permeability 38
  5.3 Factors Affecting Permeability 38
  5.4 Permeability Determination 39
      5.4.1 Permeability, Laboratory Experiment 39
      5.4.2 Permeability, Well Testing Analysis 45
      5.4.3 Permeability, Mathematical Modeling 46
      5.4.4 Permeability, Correlation 48
  5.5 Permeability Avaraging 50
  5.6 Permeability for Reservoir Simulator 54
  References 56

6 Rock Compressibility 57
  6.1 Rock Compressibility Definition and Types 57
  6.2 Stresses (Pressures) Affecting Reservoir Rock 58
  6.3 Effect of Reservoir Depletion on Pore Volume 61
  6.4 Rock Compressibility Determination 65
      6.4.1 Rock Compressibility Determination, Laboratory
            Experiments 65
      6.4.2 Rock Compressibility Determination, Correlation 67
  6.5 Rock Compressibility for Reservoir Simulator 72
  References 73

7 Wettability 75
  7.1 Wettability 75
  References 77

8 Capillary Pressure 79
  8.1 Forces Affecting Two Cotiguous Immisible Fluids 79
  8.2 Capillary Pressure Definition and Concept 81
  8.3 Capillary Pressure Determination 84
      8.3.1 Porous Plate Experiment (McCullough, 1944) 84
      8.3.2 Centrifuge Experiment (Hassler and Brunner, 1945) 85
      8.3.3 Mercury Injection Experiment (Purcell, 1949) 86
  8.4 Capillary Pressure, Lab to Rfeservoir Conversion 87
  8.5 Pore Size Distribution 88
  8.6 Free Water Level, Oil Water Contact and Transition Zone 91
  8.7 Capillary Pressure Averaging 95
      8.7.1 Leverette J-Function 95
      8.7.2 Water Saturation-Depth Profile, Pseudo Capillary
            Pressure 99
  8.8 Saturation Height Function (SHT) 99
  8.9 Reservoir Simulation Model Initialization 102
  8.10 Drainage and Imbibition Capillary Pressure Curves 107

8.11 Gas Oil Capillary Pressure Curve-Oil Reservoir and Gas
     Cap Saturation Distribution 110
8.12 Capillary Migration 111
8.13 Effects of a Pressure Gradient 113
References 114

9 Relative Permeability 117
9.1 Relative Permeability 117
9.2 Rock Typing and Rock Groups 127
9.3 Rock Properties in Reservoir Simulation 128
9.4 Saturation Functions 130
    9.4.1 Saturation Functions (Family-1) 130
9.5 Saturation Functions (Family-2) 131
9.6 Drainage and Imbibition Saturation Functions 131
9.7 Saturation Functions for Water Gas System (Gas Reservoir
    Problems) 141
9.8 Saturation Functions' End Points 144
    9.8.1 Mimum Water Saturation (Eclipse: Swl) 144
    9.8.2 Critical Water Saturation (Eclipse: Swcr) 146
    9.8.3 Water Saturation at Residual Oil (Eclipse: 1—Sowcr) 147
    9.8.4 Maximum Water Saturation (Eclipse: Swu) 147
    9.8.5 Minimum Gas Saturation (Eclipse: Sgl) 147
    9.8.6 Critical Gas Saturation (Eclipse: Sgcr) 147
    9.8.7 Gas Saturation at Residul Oil Saturation (Eclipse:
          1—Sogcr—Swl) 148
    9.8.8 Maximum Gas Saturation (Eclipse: Sgu) 148
    9.8.9 Violating End Points Rules (What Could Be
          the Effect?) 148
9.9 Preparing Saturation Functions (An Adopted Methodology) 152
9.10 Hysteresis 177
9.11 Electrical Rock Properties 180
References 190

Bibliography 191
Index 193

Chapter 1
Introduction

No doubt that the reservoir engineer mission is to maximize/optimize the hydro-carbon recovery from a given oil/gas resource. To fulfill this mission, a proper reservoir development plan is needed for any reservoir. Handling and analyzing different reservoir rock properties is an essential part in building reservoir simulation models and consequently in reservoir development plans (FDP) work. This is true because the reservoir rock is the container that holds the hydrocarbon. It is important to know the capacity of that container to hold the hydrocarbon. It is also important to understand the behavior of the container under different depletion strategies and how it affects the hydrocarbon recovery.

As everybody aware, it is a normal practice that the reservoir engineers nowadays use high technology and quite sophisticated computer programs to achieve their work. Such practice helps to ease the work, save effort and time and improve the results' accuracy. However, it may shade the basic understanding of reservoir rock properties and result in providing unrealistic data to the reservoir simulator or other used programs. This can mislead the simulator and affect its results. Within this text, the reservoir simulator is assumed to be ECLIPSE. However, handling of the reservoir rock properties by other simulators is expected to be very similar.

1.1 Trap, Reservoir, Aquifer and Basin

Before going deeply into the main articles of this text, let us revise some simple information about the hydrocarbon container. **Trap** (Fig. 1.1) is a geological structure that is able to hold fluids. In other words, it is able to prevent entering fluids from further moving. Traps were formed as a result of some structural events, some stratigraphic events or combination of the two types of events. The way any geological trap was formed suggests its type as **structural, stratigraphic** or **combined** trap (Fig. 1.2). Another interesting type of traps are the **differential traps.** This type of traps was

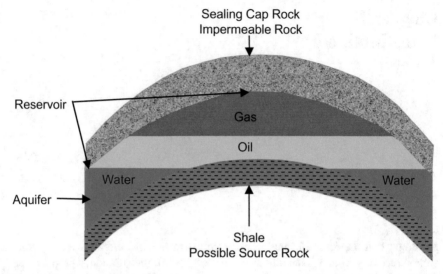

Fig. 1.1 Trap, reservoir and aquifer

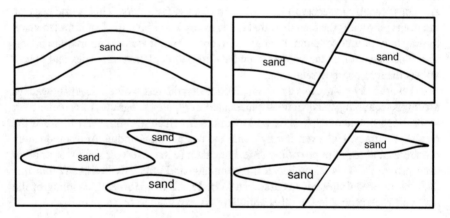

Fig. 1.2 Examples of different trap types

originally postulated theoretically but subsequently found realistic in many cases. The differential traps system consists of a series of hydraulically connected traps on different levels. The filling process end up with the heaviest fluid (water) in the shallowest trap while the lightest fluid gas occupies the deepest one (Tarek, 2020).

Hydrocarbon **reservoir** (Fig. 1.1) is the portion of the geological trap that contains oil and/or gas as a single hydraulically connected system. Within the trap, the reservoir porous body is sealed by impervious rocks or some geologic features. Figure 1.3 shows different types of traps and different types of sealing.

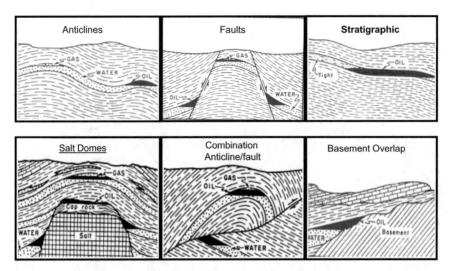

Fig. 1.3   Examples of different trap types (showing different types of sealing)

In addition to the types of traps mentioned above, another interesting type of trap known as the **differential traps** do exists. This type of trap was originally postulated theoretically but subsequently found realistic in many cases. The differential traps system consists of a series of hydraulically connected traps at different levels. Figure 1.3 shows the steps of filling different traps. The filling process end up with the heaviest fluid (water) in the shallowest trap while the lightest fluid gas occupies the deepest one (Brooks & Corey, 1964).

Many reservoirs are hydraulically connected to underlain water bearing rock known as **aquifer** (Fig. 1.1). The aquifer represents one of the driving mechanisms in reservoir depletion. Many reservoirs are located in large sedimentary **basin** and share the same aquifer (Fig. 1.4). Depletion of one reservoir in such basin will affect other reservoirs in the basin. It is not surprising to have a newly discovered reservoir whose initial reservoir pressure is less than expected by the application of regional reservoir pressure gradient. It can be simply concluded that this newly discovered reservoir is in communication or sharing the same aquifer with another producing reservoir/s (Tarek, 2020).

1.2   Modeling and Simulation

One of the main objectives of this text is to understand the usage of reservoir rock properties in reservoir simulation modeling. It is preferred that the reader should have some awareness about the subject of reservoir modeling. In this introductory section, a brief description about the concept and use of modeling and simulation

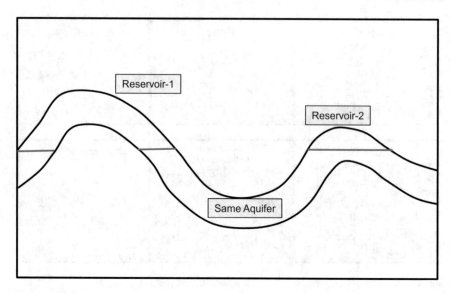

Fig. 1.4 Basin

is given. This would specially be helpful for reservoir engineers with no or limited background in reservoir simulation.

Scientific modeling is the scientific activity whose aim is to make a particular part or feature of the world easier to understand, define, quantify, visualize, or simulate. A model in science is a physical, mathematical, or logical representation of a system of entities, phenomena, or processes. In other words, a model is a simplified abstract view of the complex reality. Models are typically used when it is either impossible or impractical to create experimental conditions in which scientists can directly measure outcomes. Direct measurement of outcomes under controlled conditions will always be more accurate than model estimation. When predicting outcomes, models use some assumptions while direct measurements do not. As the number of assumptions in a model increases, the accuracy and relevance of the model decreases.

As mentioned above, the model can be physical, mathematical or logical representation of a feature. Physical modeling has been a common practice for long time. Figure 1.5 shows two physical models of two buildings. As can be recognized, the two models are of reduced scale. However, the physical model is not always smaller. Let us think about physically modeling an atom (Fig. 1.6). For such case, the model should be larger to achieve its goal.

Mathematical models represent the complex reality by one or more mathematical formula. Simple mathematical modeling has been used for quite long time. However, more sophisticated mathematical modeling has become practically applicable only with the development of digital computers.

Simulation is the imitation of the operation of a real-world process or system over time. The task of simulating something first requires that a model be developed.

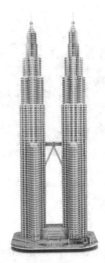

1:150 Capital theatre, Hong Kong   PETRONAS Twin Towers, Kuala Lumpur

Fig. 1.5 Building models

Carbon   Nitrogen

Fig. 1.6 Atom models

Simulation brings a model to life and shows how a particular object or phenomenon will behave. Briefly, the model represents the system itself, whereas the simulation represents the operation of the system over time. Figure 1.7 shows two examples of simulation models.

| Wooden, Mechanical 1:1 Horse Simulator (WWI) | 1:8 Live Steam Train Simulator |

Fig. 1.7 Simulators

1.3 Reservoir Simulation Modeling

Reservoir Simulation Modeling is the art of developing reliable mathematical model to be used as a tool to manage and predict reservoir performance over time. The task of reservoir simulation is to be fulfilled by following some main steps;

– Building Model
– Tuning/history matching the model
– Using the tuned/history matched model to predict reservoir performance under different depletion strategies.

Data required for building any reservoir simulation model include:

– Reservoir Geometry Data
– Reservoir Rock Properties Data
– Reservoir Fluid Properties Data
– Historical Reservoir Data (production and pressure history).

The provided data should be formulated to honor the related physical facts and the way the used simulator (software) interprets. It is thought that this part of reservoir simulation modeling activity is the most important one.

Reservoir simulation models can be categorized in different types according to different criteria. Referring to Fig. 1.8, the reservoir simulation model can be a zero dimension (0-D) model which known also as tank model, one dimension (1-D) model, two dimensional (2-D) model or three dimensional (3-D) model. The 0-D model is typically manifested in material balance studies. The other three types are related to reservoir simulation models.

On the other hand, the reservoir simulation model can be single phase (1-Ph) model, two phase (2-Ph) model or three phase (3-Ph) model. Additionally, the reservoir simulation model can be black oil model or compositional model.

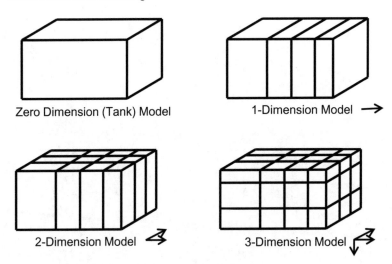

Fig. 1.8 Reservoir models

The type of the model needed for some study depend upon the nature of the reservoir and the objective of the study. The 3-Ph, 3-D, black oil model could be the most common one that serves for most required reservoir study objectives.

References

Brooks, R. H. & Corey, A. T. (1964). Hydraulic properties of porous media. (Hydraulic Paper No. 3). Colorado State University.

Tarek, G. (2020). Technical guidance for petroleum exploration and production plans. Springer International Publishing. ISBN 978–3–030–45250–6.

Chapter 2
Coring and Core Analysis

Reservoir rock properties are estimated using different approaches. These approaches include well log analysis, core analysis, well testing analysis and seismic attributes. No doubt that coring is the only way to obtain real reservoir rock samples. Consequently, core analysis is the only approach that handles real reservoir rock. In spite of its limitations, the results from core analysis should be considered to support and calibrate rock properties estimation from other approaches.

2.1 Coring, Definition and Applications

Coring is defined as the process of cutting and recovery of formation rock portion to be used for geoscience studies and measuring rock properties. The coring is done while drilling wells. It is normally performed while drilling exploration and appraisal wells. Additionally, it is performed while drilling development wells for additional information. Along the history of a reservoir, coring and core analysis are possible in stages 2–8 (see Fig.2.1).

Core analysis provides useful information that support different oil and gas industry tasks. For reservoir geology, the core analysis provides the essential information to understand reservoir rock lithology, minerals content, depositional environment and fracturing systems. For the petrophysics and reservoir engineering, the core analysis is the way to determine reservoir rock properties including porosity, permeability, fluid saturation, wettability, capillary pressure, relative permeability, electric properties and rock mechanics properties. As for drilling and well completion, core analysis provides the necessary information about fluid/formation compatibility, grain size data for gravel packing and rock mechanics data for the well completion design.

© The Author(s), under exclusive license to Springer Nature Switzerland AG 2022
A. M. Badawy and T. A. A. O. Ganat, *Rock Properties and Reservoir Engineering: A Practical View*, Petroleum Engineering,
https://doi.org/10.1007/978-3-030-87462-9_2

Fig. 2.1 Typical bottom
hole coring assembly

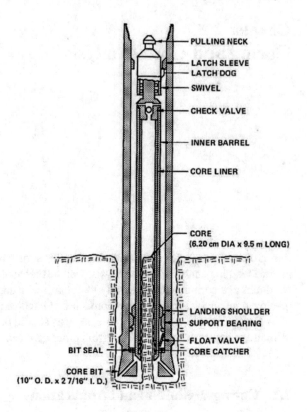

- PULLING NECK
- LATCH SLEEVE
- LATCH DOG
- SWIVEL
- CHECK VALVE
- INNER BARREL
- CORE LINER
- CORE
  (6.20 cm DIA x 9.5 m LONG)
- LANDING SHOULDER
- SUPPORT BEARING
- FLOAT VALVE
- CORE CATCHER

BIT SEAL

CORE BIT
(10" O. D. x 2 7/16" I. D.)

2.2   Types of Coring

There are different types of coring. A specific coring type is normally selected
according to the required information. Time and cost also play a role in selecting
what type of coring to select.

2.2.1   Conventional Coring

This is the most popular and maybe the most useful type of coring. A cylindrical
sample of the formation rock is cut along the axis of the drilled hole. For this purpose,
the normal drilling assembly is to be replaced by the bottom hole coring assembly
(BHCA). Figure 2.1 shows a typical bottom hole coring assembly. The hollow coring
bit (Fig. 2.2) creates a cylindrical portion of the formation rock. The cylindrical rock
sample passes through the middle of the hollow bit to be retained in the inner barrel
of the BHCA. This type of coring produces the most reliable samples for routine and
special core analysis experiments. In spite of its time consumption and high cost, the
conventional coring should be considered as the first coring priority.

Fig. 2.2 The hollow coring bit

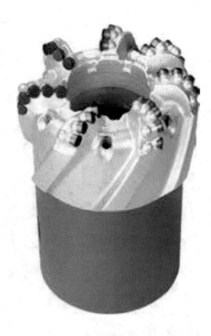

2.2.2 *Sidewall Coring*

For this coring type, plug size formation rock samples are cut perpendicular to the drilled hole axis. The special coring assembly used to cut these samples is run in hole on wire line. There are two sub-types of sidewall coring; percussion and rotary coring.

For the **percussion sidewall** coring (Fig. 2.2), a wireline tool with series of core bullets (cups) loaded with explosive charges is used. The bullets are loaded into the tool which is run in on wireline. The explosive charges shoot the bullets into formation to cut plug size samples. The mini core plugs cut by the bullets are retained within the bullets and brought back into the tool using a wire chain. This type of coring is very cheap. However, it is the least useful technique. The samples cut from hard rock can be easily fractured. The technique is unreliable for routine or special core analysis experiments. It maybe only reliable for geological description.

As for **rotary sidewall** coring, the wireline tool is equipped with a series of mini core bits. The mini core bit is extracted and pressed against the borehole wall. The bit is rotated by mud circulation through the tool to cut the plug size core sample. This coring technique produces samples which are reliable for measuring porosity and permeability but less reliable for special core analysis experiments especially for relative permeability determination. The technique is more expensive than the percussion technique but still less expensive than the conventional coring technique.

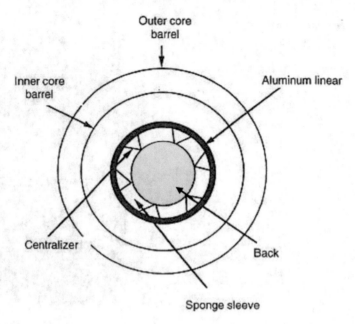

Fig. 2.3 The sponge liner

2.2.3 Special Coring Techniques

The special coring techniques are the ones that serve for special purpose coring analyses. Generally, the special coring systems are modified from the conventional coring system to produce core samples suitable for such special purposes. Additional cost is expected by using these techniques. However, it worth for the required special purpose. Three special coring techniques are discussed below.

2.2.3.1 Sponge Coring System

The inner core barrel is fitted with sponge (polyurethane). The sponge liner (Fig. 2.3) absorbs the fluids that may expel out of core during the upward trip in the case of basic conventional coring. Using this technique, the in-situ saturation can be determined. This Technique is very useful for studying enhanced oil recovery (EOR) candidate reservoirs.

2.2.3.2 Gel Coring System

This System provides a way of down hole core preservation and capsulation. The gel used for this purpose is a viscous, high molecular weight, polypropylene glycol with zero spurt loss. It is non-soluble q1 in water and environmentally safe. In the gel coring assembly, the disposal inner coring barrel is equipped with a floating rotating piston (rabbit) for gel distribution. The inner barrel is being loaded with the coring gel before coring. (about 22 gallons for 4 ¼ inch by 30 ft barrel). When the core starts to enter the coring assembly, it activates the floating piston. This causes the piston valve to be opened and the gel to flow down and distribute around the cut core. By this technique the core is immediately preserved and capsulated and consequently protected from contamination. The technique is also useful for coring unconsolidated rock.

2.2.3.3 Fiberglass Inner Coring Barrel

Fiberglass inner barrel provides lower friction core entry which makes the coring process more efficient. Fiberglass inner barrel does not react with chemicals and it should be used in high hydrogen sulfide environment. Fiberglass inner barrel can be used in high temperature wells (up to 302 °F). It can be easily cut at well site (into three feet portions) and acts as core preservation container.

2.3 Coring Fluids

The main objective of coring process is to obtain rock samples which represent the formation rock being cored. In general, the effect of mud filtrate on the formation should be minimized. To fulfil this purpose, a great care should be paid for selecting the drilling fluids used while coring (in other words, the coring fluid). Mud additives that known to be difficult to remove from the core such as some polymer types should be avoided. Additionally, mud additives that known to be able to change the rock wettability (surfactants) should be avoided.

2.4 Precautions While Coring, Core Handling and Core Transportation

To obtain reliable core at core laboratory, it is important to consider some precautions while coring, handling the recovered core at surface and transporting it to laboratory.

Firstly, we are aiming at reasonable core recovery. For this purpose, some factors should be considered during coring;

- Excessive high penetration rate (>75 ft/hr) should be avoided.
- Excessive weight on bit (>30 k.lb) should be avoided.
- The rotary speed should be controlled to be at moderate level (60–100 rpm).

- If jamming is detected during coring, it is recommended to trip out immediately. This is critical for good core recovery.
- Fast trip out should be avoided. The trip to surface should assure gradual change of temperature and pressure.

As the core reached the surface, some other precautions should be taken to avoid braking, spoiling, or damaging the core. Obviously, these precautions are more critical if dealing with fragile or unconsolidated formation;

- When laying core down to horizontal position at surface, bending core barrel should be avoided. Once horizontal, the core should not be bent or rotated.
- While still in inner barrel, the core should be marked for the well name, depth and direction (top and bottom).
- The core is to be cut into reasonable portions (normally 3 ft length portions). If disposal inner coring barrel (aluminium or fibre glass) is used, the barrel is cut with the core inside. The ends of each 3 ft portions are to be sealed with rubber or PVC caps.
- The core portions are to be kept in evacuated PVC tubes and sealed with tight cap ends. The PVC tubes may be inserted into metal tubes in case long sections are to be transported.
- The core portions are to be packed in sturdy wooden boxes for transportation.
- Dropping core tubes should be avoided when loading, unloading and sorting. Also, vibration should be avoided during transportation.

2.5 Routine and Special Core Analysis

Reservoir rock properties are determined in laboratory through experiments on small plugs (normally 1.5-inch diameter, 2.5-inch length). Those plugs are cut from the recovered whole core slab approximately at frequency of one plug at each 3 ft. It is a good practice to cut additional (twin) plugs at each plugging interval. Doing this would allow performing different experiments in parallel to save time. Additionally, this practice would secure alternative plugs that can be used in the case of damaging the essential plugs during any experiment. As mentioned previously, other popular method of getting core plugs is the sidewall coring. The plugs resulted from this process are mainly used for geological descriptive purposes. Their usage in experimental core analysis is generally unreliable. For some special purposes, some core laboratory experiments are carried out on whole core samples. The whole core used for laboratory experiments is normally a one foot cut of the full diameter core.

Core analysis experiments are divided into two major categories. Routine core analysis (RCAL) is limited with the basic reservoir rock properties such as porosity, permeability and grain density. Special core analysis (SCAL) deals with more sophisticated measurements such as rock compressibility, wettability, capillary pressure, relative permeability and rock electric properties. Due to operation concerns, it is not an easy task to have adequate core coverage for any reservoir. Then, it is important to optimize coring and core analysis program for each field. Generally speaking, the basic RCAL measurements are done for all available core plugs. The coring and core analysis programs are mainly considered in discovery wells and appraisal wells. Due to their high cost and time consumption, the SCAL experiments are being done for selected plugs only.

Chapter 3
Porosity

As previously mentioned, the concentration in this text is going to be on the reservoir rock properties which are of more interest to the reservoir engineer. The concentration will be on handling and usage in FDP work. No details will be discussed about the methods of measuring or estimating such properties. In spite of their importance in FDP work, the reservoir rock electrical properties are not of direct interest to reservoir engineers. Those properties are discussed in separate chapter at the end of this text.

3.1 Porosity and Porosity Types

Porosity could be the simplest reservoir rock property to understand and to use in reservoir studies including building reservoir simulation model. **Porosity** (Φ) is the measure of the void space within a rock. It is defined as the ratio of the volume of void space (pore volume, PV) to the bulk volume (BV) of the rock. It can be expressed as fraction or percentage.

$$\Phi = PV/BV \tag{3.1.1}$$

Porosity is a scalar dimensionless variable that tends to change in a linear manner.

The porosity that was formed during rock deposition is known as the **primary porosity**. This includes inter-granular porosity in sandstones and inter-crystalline porosity in carbonates. Additional porosity that was formed due to later geological events is considered a **secondary porosity**. It includes fractures in different rocks and solution cavities in carbonates. The secondary porosity is more popular in carbonates and basements reservoirs. For basement reservoirs the secondary porosity (fractures) is probably the only rock voids available. It is to be mentioned that a rock with dominant primary porosity is more homogeneous than the one with dominant secondary porosity (Ahmed, 2006).

© The Author(s), under exclusive license to Springer Nature Switzerland AG 2022   17
A. M. Badawy and T. A. O. Ganat, *Rock Properties and Reservoir Engineering: A Practical View*, Petroleum Engineering,
https://doi.org/10.1007/978-3-030-87462-9_3

Absolute or total porosity is the one given by the previous basic definition. **Effective or interconnected porosity** is the one calculated by replacing the volume of void space (Pore Volume, PV) by the volume of interconnected void space in the porosity basic definition.

$$\Phi_e = \text{Interconnected PV/BV} \tag{3.1.2}$$

According this definition, the effective porosity is either equal to or less than the absolute porosity. To avoid mis-use, the reservoir engineer should be aware about the definitions of total porosity and effective porosity used by some open-hole log analysts. In their analysis, the open-hole log analysts consider the total porosity as the average of the two porosity values obtained from formation density and neutron logs $[\Phi_t = (\Phi_D + \Phi_N)/2]$. This value is actually an apparent porosity value which affected by different factors especially by the water bound in shale. After doing necessary corrections to that apparent porosity value, they define the resulted value as the effective porosity (Φ_e). This is the supposed to be the correct total or active porosity and has nothing to do with interconnection.

Reservoir rock porosity is an essential parameter in determining the hydrocarbon in place and hydrocarbon reserve. The famous formulae to calculate the original hydrocarbon in place are,

$$\text{STOIIP} = [7758\,\text{GRV (N/G)} \, \Phi \, (1 - S_{wi})] / B_{oi} \tag{3.1.3}$$

$$\text{GIIP} = [7758\,\text{GRV (N/G)} \, \Phi \, (1 - S_{wi})] / B_{gi} \tag{3.1.4}$$

where,
   STOIIP = original oil in place, stb.
   GIIP = original gas in place, scf.
   GRV = gross rock volume, acre.ft.
   = A h.
   A = reservoir area, acres.
   h = average Reservoir thickness, ft.
   N/G = net to gross, ratio.
   Φ = average Reservoir Porosity, fraction.
   S_{wi} = average Initial water saturation, fraction.
   B_{oi} = initial oil formation volume factor, bbl/stb.
   B_{gi} = initial gas formation volume factor, bbl/scf.

Table 3.1 shows the range of porosity values for some types of lithology. It is to be mentioned that high porosity values do not always mean good quality rock. As can be noticed from the above table, porosity of shale can be as high as 45% but it is impermeable rock. Also, we can recognize that the vuggy limestone has higher porosity than the massive limestone. This is understood that the vuggs created by active solutions create additional secondary porosity.

Table 3.1 Porosity ranges for different lithologies

Lithology	Porosity range (%)
Unconsolidated sands	35–45
Reservoir sandstones	15–35
Compact sandstones	5–15
Shales	0–45
Massive limestones	5–10
Vuggy limestones	10–40
Dolomite	10–30
Chalk	5–40
Granite	<1
Basalt	<0.5
Conglomerate	1–15

3.2 Factors Affecting Porosity

The value of porosity can be affected by different factors. In this section, we discuss what thought to be the most important factors that affect the porosity value. **Compaction** due to overburden pressure will reduce the porosity. Accordingly, it is expected that the porosity is affected by **burial depth**. Figures 3.1 and 3.2 illus-

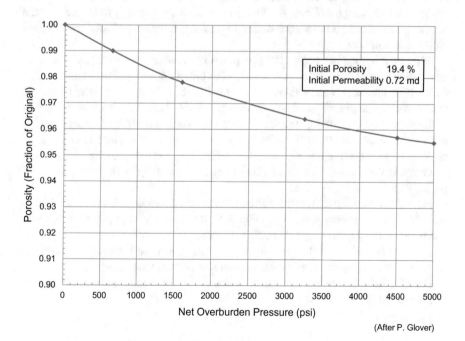

Initial Porosity   19.4 %
Initial Permeability 0.72 md

(After P. Glover)

Fig. 3.1 Porosity, effect of overburden pressure

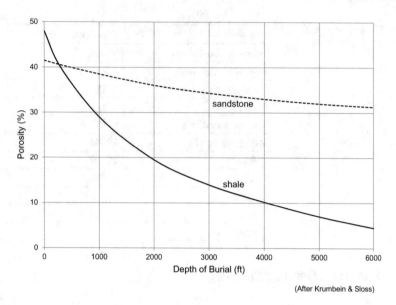

(After Krumbein & Sloss)

Fig. 3.2 Porosity, effect of burial depth

trate the effect of overburden pressure on porosity value. Overburden pressure and its effect on porosity will be more discussed when talking about reservoir rock compressibility in a later chapter (Chap. 6). When measuring the porosity in laboratory, the sample (plug) should be under pre-defined stress to simulate the effect of overburden pressure in nature.

Type of packing has great effect on porosity value. As the rock grain packing gets tighter, the porosity value gets smaller. For better understanding, let us consider a rock with identical regular spherical grains. For the cubic packing (the least tight arrangement), the porosity can be calculated as 47.6% (Fig. 3.3). Moving the grains of the upper layer one radius towards right so that each one grain of the upper layer will rest between two grains of the lower layer will result to what known as hexagonal packing and reduce the porosity to 39.5% (Fig. 3.3). Further moving of the upper layer grains one radius back (rhombhedral or close packing) will more reduce the porosity to be 26.0% (Fig. 3.3).

Rock grain **sorting** affects the porosity value. Well sorted grains rock has higher porosity than the poorly sorted one (Fig. 3.4). It is possible to say that the porosity decreases as the degree of grains sorting decreases.

Porosity decreases as the amount of **cementing and/or interstitial materials** increase. Clean sandstone has higher porosity value than shaly sandstone.

Porosity increases as the amount of **fractures and/or vugs increase**. Fractured limestone has higher porosity than the fracture free one.

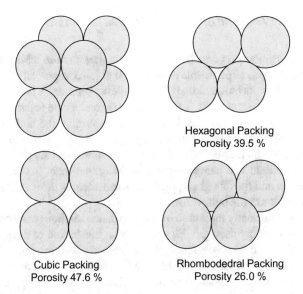

Hexagonal Packing
Porosity 39.5 %

Cubic Packing
Porosity 47.6 %

Rhombodedral Packing
Porosity 26.0 %

Fig. 3.3 Type of packing

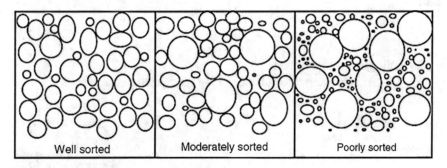

Well sorted   Moderately sorted   Poorly sorted

Fig. 3.4 Rock grain sorting

3.3 Porosity Determination

Porosity value is determined using different approaches. These approaches include visual method (core slab and thin section inspection), core measurement, open hole log analysis and seismic amplitude. Open hole log analysis is the most used approach. It can cover all interested intervals in the drilled well. The core measured porosity approach covers limited parts of the interested intervals. However, it is needed to calibrate results from open hole log.

3.3.1 Porosity Determination, Core Measurement

Porosity of rock samples (normally core plugs) is measured in laboratory as a part of routine core analysis experiments program. To obtain the porosity of a core plug, it is required to know its bulk volume (BV) and pore volume (PV). The bulk volume can be calculated mathematically if the plug has regular shape (cylinder). However, this is not normally the case. The usual way to determine the bulk volume is by immersing the plug in strongly non wetting liquid (normally mercury) or in fine glass powder. The displaced liquid or glass powder represents the plug bulk volume. It is to be commented that using mercury in this experiment is currently aborted for safety reasons and the fine glass powder displacement is the accepted method to determine the rock sample bulk volume.

Pore volume is normally measured by gas expansion method based on Boyl's law. Helium is the commonly used gas. Figure 3.5 is an illustration of the process. First, the valve connecting the two chambers is kept closed. The core plug is placed in a chamber under vacuum to remove all the air inside its pores. The other chamber is to be filled with gas under specified pressure P_1. Then the valve is opened to allow the gas to flow to the plug pores and the new pressure P_2 is to be recorded. Naturally P_2 will be less than P_1. Finally, the plug pore volume can be calculated applying Boyl's law,

$$P_2 V_2 = P_1 V_1 \qquad\qquad (3.3.1.1)$$

V_1 = the gas chamber volume.
$V_2 = V_1 +$ connecting tube volume + Plug Pore Volume chamber.

Fig. 3.5 Schematic diagram of double cell boyle's law porosimeter

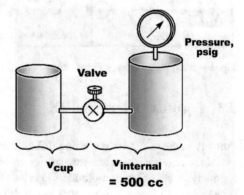

3.3.2 Porosity Determination, Open Hole Logs

Open hole logs analysis is the routine approach for determining the porosity. Using this approach, a good porosity distribution can be established all over the reservoir. This is because, open hole logging is a basic data collection technique while drilling any well. However, porosity obtained using this approach should be calibrated with the porosity obtained from core measurement. The porosity logging tools include sonic log, formation density log and neutron log. It is important to note that these porosity tools are shallow investigation depth. They mainly affected by the fluids in the invaded zone. In the case of clean formation with known lithology and absence of secondary porosity, either tool is enough to give good porosity estimation. Generally, this is not the case and combination of more than one tool is necessary. Fortunately, each tool responds differently to formation lithology and its fluid content. By skillful interpretation, the combination of different tools can give good porosity estimation as well as formation lithology components.

Sonic log porosity estimation is based on the different sound wave travel velocity in different media. Interval transit time (Δt) for a formation is defined as the time required by sound wave to travel one foot through this formation. The unit used to define the interval transit time is $\mu s/ft$. A transmitter in the sonic log tool generates sound waves which travel through the formation in the vicinity of the well bore. Then, a receiver in the tool detects these sound waves upon their return. The time lapse between generation and detection is converted to interval transit time and recoded continuously versus the depth of the tool. The porosity versus depth can then be calculated by,

$$\Delta t_{log} = \Phi \Delta t_f + (1 - \Phi) \Delta t_{ma}$$

or $\hspace{5cm}$ (3.3.2.1)

$$\Phi = \left(\Delta t_{10g} - \Delta t_{ma} \right) / \left(\Delta t_f - \Delta t_{ma} \right)$$

where,

Φ = porosity, fraction.

Δt_{log} = formation interval transit time as recorded by the tool, $\mu s/ft$.

Δt_{ma} = rock matrix interval transit time, $\mu s/ft$.

Δt_f = fluid interval transit time, $\mu s/ft$.

It is believed that the porosity determined by sonic log represents the primary porosity only. This is because the sound waves ignore the fractures and vuggs and travel only through the rock matrix and intergranular pores. For clean carbonate formation, the **secondary porosity index** ($I_{\Phi 2}$) is calculated as the difference between porosity from density and/or neutron logs (Φ_{DN}) and the one from sonic log (Φ_S) (Table 3.2),

$$I_{\Phi 2} = \Phi_{DN} - \Phi_S (1 - \Phi)$$ $\hspace{3cm}$ (3.3.2.2)

Table 3.2 Shows values of sound travel time of main formation rock materials

Lithology	Δt (μs/ft)	ρ (gm/ cm³)
Sandstone	55.5	2.65
Limestone	47.5	2.71
Dolomite	43.5	2.87
Anhydrite	50.0	2.96
Rock salt	67.0	2.17
Water	189.0	1.00

Formation density log tool is a type of nuclear tool. Radioactive source in the tool emits medium energy gamma ray into the formation. This gamma ray interacts with the electrons in the formation and loose some of its energy. At some level of lower energy, it can be partially absorbed by the electrons. The attenuated flux of gamma ray that reaches sensors in the tool reflects the amount of attenuation which depends upon the density of electrons in the formation. Formation bulk density is directly related to electron density and can be estimated and recorded versus depth. Knowing formation bulk density, the formation porosity can be calculated by,

$$\rho_b = \Phi\rho_f + (1 - \Phi)\rho_{ma}$$

or   (3.3.2.2)

$$\Phi = (\rho_{ma} - \rho_b)/(\rho_{ma} - \rho_f)n$$

where,

Φ = porosity, fraction.

ρ_b = formation bulk density as recorded by the tool, gm/ cm³.

ρ_{ma} = rock matrix density, gm/ cm³.

ρ_f = fluid density, gm/ cm³.

Neutron log tool is another type of nuclear tool. High energy neutrons are emitted into the formation from a radioactive source. These fast neutrons react with the nuclei of the atoms in the formation. At each reaction (collision with nucleus), a high energy neutron loses some energy and slows down while the nucleus gains some energy. This energy transfer is more efficient when the masses of slowed down neutron and the nucleus within the formation have the almost the same mass. In other words, the energy transfer is more efficient when a neutron collides with hydrogen nucleus. The collision continues and the neutrons lose more energy and a number of them can finally be absorbed by the formation atoms. The tool detector which is placed some distance from the radioactive source measures the population of the absorbed neutrons, the population of the remaining neutrons or the gamma ray emitted when neutrons are absorbed. This measurement reflects the amount of hydrogen in the formation. As hydrogen is fundamentally associated to the amount of water and/or hydrocarbon present in the pore space, the detector measurement is directly linked to porosity (Schlumberger, 1987).

3.4 Porosity Averaging

Porosity is a scalar non dimension quantity that can be simply averaged using arithmetic or weighted arithmetic averaging method. Considering number of layers with different porosity and different thickness values (Fig. 3.6), the average porosity (Φ_{avg}) would be calculated as the weighted arithmetic average of the individual layers' porosities. The weight used is the individual layer thickness.

$$\Phi_{axg} = \Sigma(\Phi_i * h_i)/\Sigma(h_i), i = 1 \text{ to } n \tag{3.4.1}$$

It is obvious that in the case of same thickness layers, the simple arithmetic average can be applied;

$$\Phi_{axg} = \Sigma(\Phi_i)/n \tag{3.4.2}$$

where,

Φ_{avg} = average porosity.
Φ_i = individual layer porosity.
h_i = individual layer thickness.
n = number of layers.

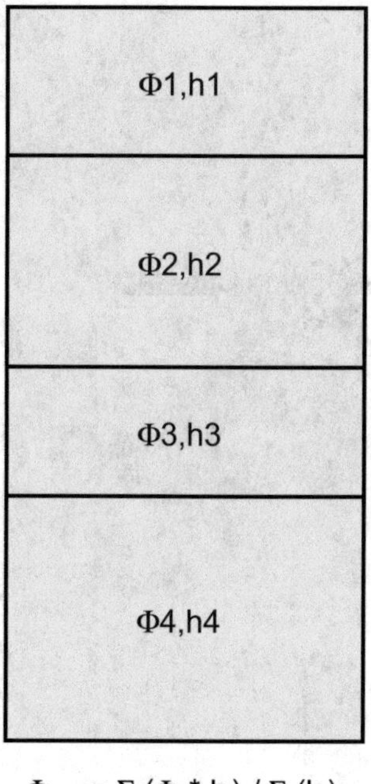

Fig. 3.6 Different thickness layers

$$\Phi_{avg} = \Sigma(\Phi_i * h_i) / \Sigma(h_i)$$

In reservoir modeling work, it is needed to up-scale from fine blocks model to coarser blocks model. For the up-scaling procedure, number of fine blocks are being combined to one coarser block. Figure 3.7 shows the case of having identical same size fine blocks. For such similar blocks, simple arithmetic average (Eq. 3.4.2.), using n as the number of fine blocks, should be good enough to estimate the up-scaled coarser block average porosity. However, in the case of fine blocks are not similar (Fig. 3.8), an appropriate weighting factor should be considered. Normally the block bulk volume (BV) is used as a weighting factor to calculate weighted arithmetic average porosity.

$$\Phi_{\text{ayg}} = \Sigma(\Phi_i * BV_i)/\Sigma(BV_i), i = 1 \text{ to } n \tag{3.4.3}$$

Fig. 3.7 Identical blocks

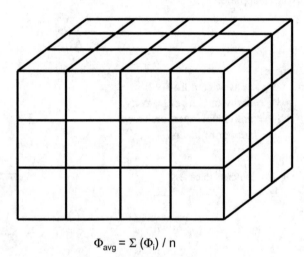

$$\Phi_{\text{avg}} = \Sigma(\Phi_i)/n$$

Fig. 3.8 Non-Identical blocks

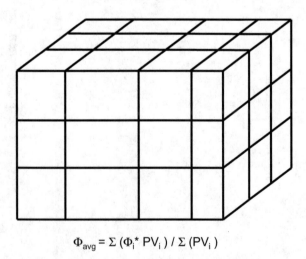

$$\Phi_{\text{avg}} = \Sigma(\Phi_i * PV_i)/\Sigma(PV_i)$$

where,

Φ_{avg} = average coarser block porosity.
Φ_i = individual fine block porosity.
BV_i = individual fine block bulk volume.
n = number of fine blocks.

3.5 Porosity for Reservoir Simulator

Porosity is introduced to reservoir simulator as fractions in array form. One porosity value for each model block. The values given must be equal to or greater than zero. If zero porosity set to some blocks, the simulator will consider these blocks as inactive and will not consider any in or out flow for them. The porosity array is normally imported from the output of the geological model. Here in is a simple example of porosity introduced to simulator for a hundred blocks one dimensional reservoir model.

PORO

```
0.15 0.15 0.15 0.15 0.15 0.15 0.15 0.15 0.15 0.15
0.15 0.15 0.15 0.15 0.15 0.15 0.15 0.15 0.15 0.15
0.18 0.18 0.18 0.18 0.18 0.18 0.18 0.18 0.18 0.18
0.18 0.18 0.18 0.18 0.18 0.18 0.18 0.18 0.18 0.18
0.18 0.18 0.18 0.18 0.18 0.18 0.18 0.18 0.18 0.18
0.20 0.20 0.20 0.20 0.20 0.20 0.20 0.20 0.20 0.20
0.20 0.20 0.20 0.20 0.20 0.20 0.20 0.20 0.20 0.20

0.20 0.20 0.20 0.20 0.20 0.20 0.20 0.20 0.20 0.20
0.21 0.21 0.21 0.21 0.21 0.21 0.21 0.21 0.21 0.21
0.21 0.21 0.21 0.21 0.21 0.21 0.21 0.21 0.21 0.21
```

or simpler by making use of the repeated values,

PORO

```
20*0.15   30*0.18   30*0.20   20*0.21
```

Exercises

1   Prove that the porosity of a rock consists of similar spherical grains with cubic packing is equal to 47.6%
2   A core plug has regular cylindrical shape with 2.5 cm diameter and 5.0 cm length. The plug was used for porosity determination. The pore volume was estimated using gas expansion method. The gas chamber volume is 20 cm^3. Initial pressure

of the experiment was recorded as 30 psig while the ending pressure was recoded as 20 psig. Ignoring the connecting tube volume, calculate the porosity of the core plug.

3   Density log and sonic log were run in a discovery well. At some depth, the formation bulk density was recorded as 2.45 gm/cm^3 and the interval transit time was recorded as 65.0 · s/ft. The mud filtrate density is 1.00 gm/cm^3. If the formation lithology at that depth is known to be clean limestone, calculate the formation porosity as given by the two tools. Compare the two porosity values and explain the possible reason of being different.

4   An oil reservoir consists of four layers with properties as shown in the table:

Layer number	1	2	3	4
Thickness (ft)	3	5	2	7
Porosity (%)	20	15	10	17

Calculate the weighted average reservoir porosity.

If the all the layers have the same thickness, what would be the average reservoir porosity?

Reference

Ahmed, T. (2006). *Reservoir Engineering Handbook, 3rd Edition*. Gulf Professional Publishing.
Schlumberger (1987). Log Interpretation Principles/Applications.

Chapter 4
Fluid Saturation

4.1 Fluid Saturation, Importance and Types

Fluid saturation is the measure of the void space within a rock filled with a specific fluid. It is defined as the ratio of the pore volume filled with a specific reservoir fluid to the total pore volume (PV). Accordingly, water saturation (S_w) is defined as,

$$S_w = \text{Water Filled PV / Total PV} \tag{4.1.1}$$

Similar formulae can be written for other fluids' saturation in the reservoir system; oil saturation (S_o) and gas saturation (S_o).

It is obvious that the saturations of all fluids in the system are summed up to unity;

$$S_w + S_0 + S_g = 1.0 \tag{4.1.2}$$

Like porosity, saturation is a dimensionless scalar variable. It can be expressed as fraction or percentage. Fluid saturation is one of the essential parameters for estimating the hydrocarbon in place. It is also needed for studying fluid flow in reservoir and reservoir development planning.

In hydrocarbon reservoir studies, we should be clear about different types of the fluids' saturation. The following discussion elaborates about these types.

Connate Water Saturation (S_{wc}) is defined as the water saturation at time of reservoir discovery. It can have any value from the irreducible water saturation to the maximum of 100 value 3.0–5.0%)% in the aquifer. Connate water (also called fossil water) was trapped in the pores of the rock during its formation. The chemistry of the connate water could change along the rock history.

Irreducible water saturation (S_{wirr}) is the minimum possible value of water saturation in the reservoir system. However, it is generally determined in laboratory and not always detected in reservoir. Its value depends upon the rock quality.

© The Author(s), under exclusive license to Springer Nature Switzerland AG 2022
A. M. Badawy and T. A. A. O. Ganat, *Rock Properties and Reservoir Engineering: A Practical View*, Petroleum Engineering,
https://doi.org/10.1007/978-3-030-87462-9_4

Critical Water Saturation (S_{wcr}) is the minimum water saturation at which the water phase starts to be mobile. In most applications, it is considered equals to the irreducible water saturation.

At any point in an oil reservoir, the initial oil saturation is calculated by,

$$S_0 = 1.0 - S_w \qquad (4.1.3)$$

Similar equation can be written for gas reservoir,

$$S_g = 1.0 - S_w \qquad (4.1.4)$$

Residual Oil Saturation (S_{or}) is the oil saturation at which the oil phase become immobile and it is the minimum possible oil saturation in the oil reservoir. This saturation is reached as result of complete displacement of oil from the pores. There are two different residual oil saturation, the water–oil residual oil saturation (S_{orw}) which reached by water displacing oil and the gas-oil residual saturation (S_{org}) caused by gas displacing oil. These residual oil saturations are the ones that can be achieved by primary drive forces and secondary recovery methods. Successful tertiary recovery methods would reduce the residual oil saturation and consequently increase ultimate oil recovery. We should be clear about the difference between residual oil saturation and remaining oil saturation.

Remaining Oil Saturation (ROS) is the average oil saturation in a reservoir at the end of its production life. When an oil reservoir comes to its economic limit, there will be pores that still have mobile oil.

Critical Gas Saturation (S_{gcr}) is the minimum free gas saturation at which the gas phase starts to be mobile. The value of critical gas saturation is generally small (typically 3.0 to 5.0%).

Residual Gas Saturation (S_{gr}) in gas reservoir is similar to residual oil saturation in oil reservoir.

4.2   Fluid Saturation Determination

When talking about determining saturation, it is by default means determining water saturation. At initial reservoir condition, it is essential to have good water saturation estimation to calculate the initial fluid in place. Later in reservoir life, information about current saturation is important for updating reservoir development schemes especially where enhanced oil recovery is considered.

Fluid saturation is determined in laboratory as a part of routine core analysis (RCAL). This is what defined as the direct fluid saturation determination. Fluid saturation can also be estimated as a result of open-hole log interpretation which defined as the indirect fluid saturation determination. The open-hole log saturation determination is the more common due to its simplicity and availability.

Fig. 4.1 Different thickness layers

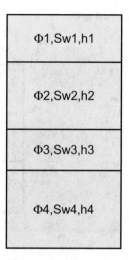

$$\text{Average } S_w = \Sigma \ (\Phi_i * h_i * S_{wi}) \ / \ \Sigma \ (\Phi_i * h_i)$$

4.3 Fluid Saturation Averaging

Average fluid saturation is estimated almost in similar way like porosity. For the case of layers with different thickness, different porosity and different water saturation values (Fig. 4.1), the average water saturation is evaluated using;

$$\text{Average } S_w = \Sigma(\Phi_i, h_i S_{wi})/\Sigma(\Phi_i, h_i), i = 1 \text{ to } n \text{ (number of layers)} \quad (4.3.1)$$

The product (ϕh) is used as the weighting factor for this weighted artithmetaic averaging formula. For equal thickness layers, the thickness value (h) can simply be eliminated from the weighting factor.

For up-scaling identical blocks (Fig. 4.2) and non identical blocks (Fig. 4.3), formulae (4.3.2) and (4.3.3) are used for calculating the average water saturaion;

$$\text{Average } S_x = \Sigma(\Phi_i S_{wi})/\Sigma(\Phi_i), i = 1 \text{ to } n \text{ (number of blocks)} \quad (4.3.2)$$

$$\text{Average } S_x = \Sigma(P V_i S_{wi})/\Sigma(P V_i), i = 1 \text{ to } n \text{ (number of blocks)} \quad (4.3.3)$$

4.4 Fluid Saturation for Reservoir Simulator

By default, the reservoir simulator calculates the initial water saturation for each reservoir block using the introduced initial pressure information, fluids' pressure gradients and capillary pressure data. This option is known as the **hydrostatic equilibrium option**. The use of this default option assures the initial stability of the

Fig. 4.2 Identical blocks

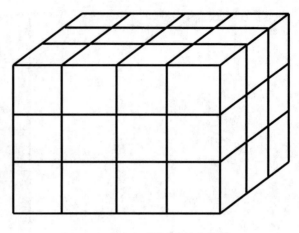

Average $S_w = \Sigma\,(\Phi_i * S_{wi})\,/\,\Sigma(\Phi_i)$

Fig. 4.3 Non-Identical blocks

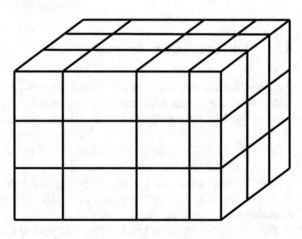

Average $S_w = \Sigma(PV_i * S_{wi})\,/\,\Sigma(Pv_i)$

model. More details about model initialization with hydrostatic equilibrium will be given in Chap. 8 (capillary pressure).

In some cases, it is preferred to provide initial fluid saturation explicitly. This is known as the **enumeration or explicit option**. The Initial water saturation is estimated out-side (normally within static model work activities). Then, the estimated values are introduced to the reservoir simulator in array form using SWAT key word. To fulfill the reservoir model initial condition, either initial oil saturation or initial gas saturation should also be provided.

Here in is a simple example of initial water and oil saturations introduced to simulator for a hundred blocks one dimensional reservoir model,

SWAT

1.000 1.000 1.000 1.000 1.000 1.000 1.000 1.000 0.993 0.850
0.733 0.647 0.558 0.487 0.429 0.405 0.381 0.376 0.365 0.355
0.348 0.345 0.337 0.332 0.326 0.323 0.321 0.316 0.313 0.311
0.309 0.307 0.305 0.303 0.301 0.299 0.297 0.295 0.293 0.291
0.289 0.287 0.286 0.285 0.284 0.283 0.282 0.281 0.280 0.279
0.278 0.277 0.276 0.275 0.274 0.273 0.272 0.271 0.270 0.269
0.268 0.267 0.266 0.265 0.264 0.264 0.263 0.263 0.262 0.262
0.261 0.261 0.260 0.260 0.259 0.259 0.258 0.258 0.257 0.257
0.256 0.255 0.255 0.254 0.254 0.253 0.253 0.253 0.252 0.252
0.252 0.251 0.251 0.251 0.251 0.250 0.250 0.250 0.250 0.250

SOIL

0.000 0.000 0.000 0.000 0.000 0.000 0.000 0.000 0.007 0.150
0.267 0.353 0.442 0.513 0.571 0.595 0.619 0.624 0.635 0.645
0.652 0.655 0.663 0.668 0.674 0.677 0.679 0.684 0.687 0.689
0.691 0.693 0.695 0.697 0.699 0.701 0.703 0.705 0.707 0.709
0.711 0.713 0.714 0.715 0.716 0.717 0.718 0.719 0.720 0.721
0.722 0.723 0.724 0.725 0.726 0.727 0.728 0.729 0.730 0.731
0.732 0.733 0.734 0.735 0.736 0.736 0.737 0.737 0.738 0.738
0.739 0.739 0.740 0.740 0.741 0.741 0.742 0.742 0.743 0.743
0.744 0.745 0.745 0.746 0.746 0.747 0.747 0.747 0.748 0.748
0.748 0.749 0.749 0.749 0.749 0.760 0.750 0.750 0.750 0.750

Exercises

1   A core plug has regular cylindrical shape with 2.5 cm diameter and 5.0 cm length.
    The plug was used for porosity determination. The pore volume was estimated
    using gas expansion method. The gas chamber volume is 20 cm^3. Initial pressure
    of the experiment was recorded as 30 psig while the ending pressure was recoded
    as 20 psig. Ignoring the connecting tube volume, calculate the porosity of the
    core plug.

2   An oil reservoir consists of four layers with properties as shown in the table:

Layer number	1	2	3	4
Thickness (ft)	3	5	2	7
Porosity (%)	20	15	10	17
Water saturation (%)	25	27	38	57

Calculate the weighted average water saturation.

If the all the layers have the same thickness, what would be the average water saturation?

Chapter 5
Permeability

5.1 Permeability, Darcy's Law

Permeability is the measure of the porous medium capacity to transit fluids. It was introduced for the first time through the experimental work by Darcy (1856). His study of the water movement through sand in water purification system led to establishing the famous formula (Darcy's Law). This law relates the fluid flow through the porous medium with the applied differential pressure. Figure 5.1 is a schematic representation of Darcy's experiment.

Considering a horizontal cylindrical sample of a porous medium (Fig. 5.2), Darcy's Law, in c.g.s. units, will be written as,

$$Q = (KA/\mu)\,(\Delta P/\Delta L) \qquad\qquad (5.1.1)$$

where,

Q = Flow rate, cc/s
A = Cross sectional area of the sample, cm^2
μ = Fluid Viscosity, poises
ΔP = Pressure Difference $(P_1 - P_2)$, dyne/ cm^2
ΔL = Length of the sample, cm
K = Constant (Later defined as the porous medium permeability).

Dimension analysis of Darcy's law shows that the permeability (K) has the dimension of area. Accordingly, it is possible to be represented in any area unit. In the original Darcy's work, the permeability was defined in the units of square centimeter. However, this unit was found to be too large for practical purposes. The industry adopted permeability unit is the darcy (1 darcy = 9.869×10^{-9} square centimeter or 1 darcy = 1.062×10^{-11} square foot). The **darcy** is defined as the permeability of a porous medium which is 100% saturated with a single fluid of 1.0 centipoise viscosity when it allows the viscous flow of that fluid at a rate of 1.0 cubic centimeter per second

Fig. 5.1 Schematic representation of Darcy's experiment

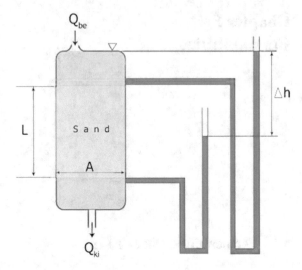

Fig. 5.2 One direction flow, Darcy's law

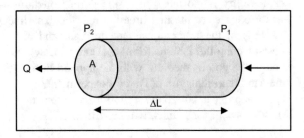

per square centimeter under a pressure gradient of 1.0 atmosphere per centimeter. The more common practical unit in industry is the millidarcy (0.001darcy).

Darcy's law can be re-written, in practical field units, to calculate single phase linear flow (Fig. 5.3) as,

$$Q = 0.00127(K * A/\mu)\,(\Delta P/\Delta L) \tag{5.1.2}$$

Fig. 5.3 Ideal linear flow

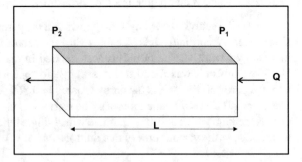

where,

  Q = Flow rate, bbl/day
  A = Cross sectional area for flow, ft^2
  μ = Fluid Viscosity, centipoises
  $\Delta P/\Delta L$ = Pressure Gradient, psi/ft
  K = Permeability, md.

For radial flow (Fig. 5.4), Darcy's law in practical field unit is written as,

$$Q = 0.007082(Kh/\mu)[(P_e - P_w)/\ln(r_e/r_w)] \tag{5.1.3}$$

where,

  Q = Flow rate, bbl/day
  h = Net Pay Thickness, ft^2
  μ = Fluid Viscosity, centipoises
  P_e = External Pressure (Reservoir Pressure, psi
  P_w = Internal Pressure (BHFP), psi
  r_e = External Radius (Drainage Radius), ft
  r_w = Internal Pressure (Borehole Radius), ft
  K = Permeability, md.

Fig. 5.4 Ideal radial flow into a wellbore

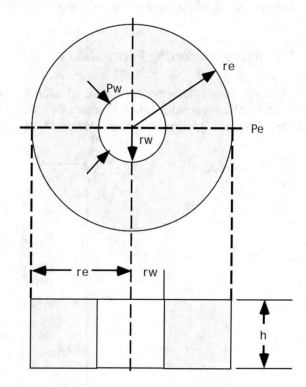

5.2 Absolute, Efective and Relative Permeability

The permeability of the porous medium when it is 100% saturated with a single fluid is defined as the **absolute permeability**. This absolute permeability (K_{abs}) is the same regardless of which fluid saturates the rock. When saturated with more than one fluid, the same porous medium has different permeability values to each specific fluid. The permeability of a porous medium to one specific fluid in the existence of other fluid/s is defined as the **effective permeability** to that fluid. A porous medium that contains oil, water and gas has effective permeability to each one of the three phases (K_o, K_w and K_g). It is important to note that ($K_o + K_w + K_g <$ K_{abs}). It is also important to insist that the permeability is a rock property. It is quite misleading to say air permeability or oil permeability. This should be replaced by permeability to air or permeability to oil. **Relative permeability** to one specific fluid is the ratio of the effective permeability to that fluid to the absolute permeability of the rock. It was found that the relative permeability to a specific fluid changes with its relative saturation. The issues related to relative permeability will be discussed in more details in Chap. 9. Permeability is a vector variable which has different values in different directions (Fig. 5.5). It tends to change in logarithmic or power manner. It is a common practice to consider two different permeability values in; horizontal permeability ($K_h = K_x = K_y$) and vertical permeability ($K_v = K_z$). However, such simplification should be avoided for reservoirs with high level of anisotropy.

5.3 Factors Affecting Permeability

It is natural that all the factors that mentioned before to affect the porosity values will also affect the permeability values. However, the effect on permeability can be more. This is because the permeability tends to change in logarithmic or power way. This is

Fig. 5.5 Permeability is a vector variable

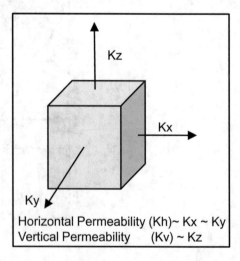

Horizontal Permeability (Kh)~ Kx ~ Ky
Vertical Permeability   (Kv) ~ Kz

especially recognized with the existence of fractures and/or vugs in carbonate reservoir rock. The existence of such fractures and/or vugs could dramatically increase the permeability value.

5.4   Permeability Determination

For a reservoir study, permeability can be estimated using different approaches. These approaches include laboratory experiment, well testing, open-hole logs and mathematical modeling. Laboratory measurement is the basic approach. Well testing permeability estimation approach is used as matching and supporting approach. Other approaches are not very popular and may be considered in the absence of laboratory experiments.

5.4.1   Permeability, Laboratory Experiment

Permeability is measured in laboratory as a part of RCAL work program. The experiment is achieved by allowing the flow of some fluid through core sample under controlled differential pressure. Normally air or some inert gas is used as the flowing fluid. Measuring permeability in laboratory is affected by different factors and special precaution should be considered. These factors are discussed below.

5.4.1.1   Net Overburden Pressure

The value of permeability is affected by the compaction due to overburden pressure (Fig. 5.6). Increasing confining (net or effective overburden) pressure reduces the porous medium pore throat and subsequently reduces the absolute permeability. When measuring the permeability in laboratory, the sample should be under predefined stress to simulate the effect of overburden pressure in nature (Tarek, 2019).

5.4.1.2   Reactive Fluids

The reactive fluid is another factor that affects the value of permeability (Fig. 5.7). For reservoir rock containing water sensitive clay, less saline water may cause clay swelling and reduce the rock permeability. Additionally, the aqueous phase may loosen the clay particles. Those particles would move and bridge at small cross sections causing flow restriction (permeability reduction). The drilling fluids should be carefully chosen to avoid altering permeability values. On the other hand, in-active fluids should be used in laboratory experiment.

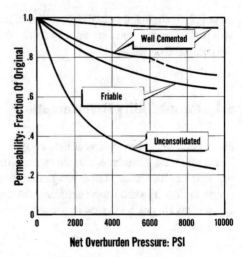

Fig. 5.6 Permeability reduction with overburden pressure (CoreLab, 1983)

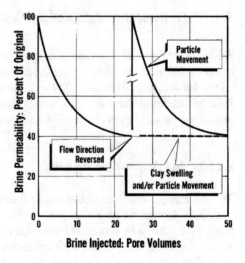

Fig. 5.7 Permeability reduction due to ClaySwelling and/or particle movement (CoreLab, 1983)

5.4.1.3 Klinkenberg Effect

As mentioned previously, air or some inert gas is used as the flowing fluid to measure permeability in laboratory. The use of gas in these experiments may result in estimating permeability greater than the real absolute permeability of the porous medium. The issue was first highlighted by Muskat (1937). This effect was further studied by Klinkenberg (1941) who interpreted it to be due to slip flow between gas molecules and solid rock material walls. In other words, this Klinkenberg effect is caused by the **gas slippage** phenomenon.

Before going in more details about Klinkenberg effect, let us have few words about the gas slippage phenomenon. Considering fluid (liquid/gas) flow in capillary tube (simulating flow in porous medium), it can be noticed that the velocity profile of liquid is maximum at the center of the tube and zero at its walls. However, the velocity profile of gas is not zero at the walls (Fig. 5.8). During travelling through the porous medium, the gas molecules are in continuous motion and collide to each other and to the pore walls of this porous medium. The collision with the pore walls leads to additional flow of gas molecules (slip flow) not due to pressure differential. This additional flow, in turn, will lead to estimating higher permeability for the porous medium (Wu et al., 1998). The amount of this additional flow (slip flow) is affected by the porous medium pore size and the mean pressure of the system. To understand these effects, we need first to look at the definition of the gas mean free path. The **gas mean free path** is the average smallest distance that a gas molecule travels before colliding with other molecules. Where the pore size is smaller and gets closer to the gas mean free path, the chance of gas molecules collision to the pore walls increases and consequently the amount of slip flow increases. This simply leads to increasing Klinkenberg effect with decreasing pore size or in other words decreasing permeability. It has been found that the Klinkenberg effect is not significant for permeability above 100 md. To understand the effect of mean pressure on Klinkenberg effect, we need to remember that the gas is very compressible substance. At higher pressures the gas is compressed, the mean free path decreases and the chance of gas molecules collision to the walls of the pore decreases as well. It is then concluded that Klinkenberg effect decreases as the mean pressure increases. As pressure gets higher and higher the gas is more compressed and its flow profile approaches the liquid flow profile. Simply, it can be stated that Klinkenberg effect is only important at low pressure (laboratory experiment) and insignificant at high pressures (reservoir condition).

Figure 5.9 shows that the permeability estimated using a specific gas is a linear function of the reciprocal of the mean pressure $(1/P_m)$ applied during the experiment.

$$K_{gas} = K_{abs} + C(1/P_m) \qquad (5.4.1.3.1)$$

where,

K_{gas} = measured permeability to gas used in the experiment, md
K_{gas} = equivalent permeability to liquid or absolute permeability, md
P_m = mean pressure applied in the experience, psi.

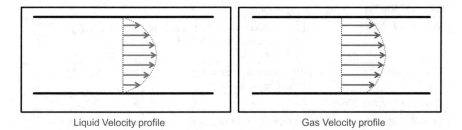

Liquid Velocity profile   Gas Velocity profile

Fig. 5.8 Fluid velocity profiles in capillary tube

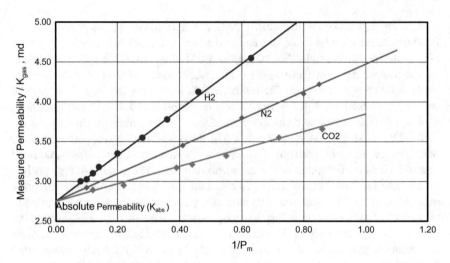

Fig. 5.9 Klinkenberg effect

It can be recognized from the same plot, that the linear relationship predicts the absolute permeability at $1/P_m$ equals to zero. Consequently, at least 3 laboratory measurements with 3 different mean pressures are needed to establish the linear relationship and estimate the absolute permeability.

The constant C in the above formula was found to be a function of the absolute permeability of the rock, the type of gas used in the experiment and the average radius of rock capillaries (Klinkenberg, 1941). Equation 5.4.1.3.1 can be re-written as,

$$K_{gas} = K_{abs}(1 + b/P_m) \qquad (5.4.1.3.2)$$

In this equation, b is defined as the **Klinkenberg Slip Factor**. Table 5.1 shows different correlations to estimate this factor.

Jones (1972) correlation looks to be the most popular one. He worked on about 100 samples with absolute permeability ranges from 10^{-17} to 10^{-12} m^2 (0.01–1000

Table 5.1 Klinkenberg slip factor correlation

Heid et al. (1950)	11 Synthetic cores and 164 natural cores from different producing fields in USA	$b = 11.419(K_{abs})^{-0.39}$
Jones (1970)	100 samples, wide range of absolute permeability	$b = 6.9(K_{abs})^{-0.36}$
Jones and Owens (1979)	–	$b = 12.639(K_{abs})^{-0.33}$
Sampatha and Keighen (1981)	10 core samples from a tight gas sand	$b = 13.851(K_{abs}/\Phi)^{-0.53}$

Table 5.2 N-Raphson iterative method

K_i	$f(K_i)$	$f'(K_i)$	K_{i+1}
K_1	$f(K_1)$	$f'(K_i)$	K_2
K_2	$f(K_2)$	$f'(K_2)$	K_3
K_3	$f(K_3)$	$f'(K_3)$	K_4

Continue iteration to find $f(K_i) = 0.0$ or $K_{i+1} = K_i$

md) and used air as the flowing gas to correlate the b (Klinkenberg slip factor) with the absolute permeability. As a result, the Eq. 5.4.1.3.2 can be rewritten as,

$$6.9(K_{abs})^{0.64} + P_m K_{abs} - P_m K_{gas} = 0 \qquad (5.4.1.3.3)$$

This equation can be solved for the value of K_{abs} using **Newton–Raphson** iterative method. It is used to estimate the absolute permeability in the case of having a single laboratory measurement. It is to be noted that this equation is only applicable when air is used as the flowing fluid in the experiment. Similar equations can be derived for other correlations.

To elaborate on Newton–Raphson iterative method, let us consider the left-hand side of the above Eq. (5.4.1.3.3) and follow the following steps (Table 5.2):

(1)   Re-write as a function of K_{abs},

$$f(K_{abs}) = 6.9(K_{abs})^{0.64} + P_m K_{abs} - P_m K_{gas}$$

   or simply,

$$f(K) = 6.9(K)^{0.64} + P_m K - P_m K_{gas}$$

(2)   Take derivatives for both sides,

$$f'(K) = 4.416(K)^{-0.36} + P_m$$

   A specific value of K (K_{i+1}) can be estimated knowing the previous guess of K (K_i),

$$K_{i+1} = K_i - f(K_i)/f'(K_i)$$

(3)   Assume a first guess of K as K1 and follow steps in Table 5.2 to give
(4)   the best estimation for K.

   For Example,
   If $P_m = 15$ psi and $K_{gas} = 103$ md.
   Then,

Table 5.3 Newton–Raphson
iterative method

K_i	f (K)	f'(K)	K_{i+1}
100.0000	86.476790	15.84145	94.5411
94.5411	−0.046289	15.85863	94.5440
94.5440	0.000000	15.85862	94.5440

$$f(K) = 6.9(K)^{0.64} + P_m K - P_m K_{gas}$$
$$f'(K) = 4.416(K)^{-0.36} + P_m$$
$$K_{i+1} = K_i - f(K_i)/f'(K_i)$$
$$K_1 = 100 \text{ md(first guess)}$$
$$f(K) = 86.47679$$
$$f'(K) = 15.84145$$

$K_2 = 94.5411$ and so on………
Refer to Table 5.3.

5.4.1.4 Permeability Measurement for Carbonate Reservoirs

Due to extreme heterogeneity characterizing vuggy carbonate, it is very possible
that the used plugs are not really representing the flow capacity of the reservoir rock.
The core sample shown in Fig. 5.10 was taken from Silurian reef. The measured
permeability for a plug cut from this sample is less than 0.1 md. Obviously, this is a
dramatic underestimation of this sample permeability. Using this measured permeability in any reservoir study will be quite mis-leading. In carbonate reservoir, the core
sample must be visually examined before using the plug measured permeability. It is
recommended to verify the core plug measured permeability with the one evaluated
from well test analysis.

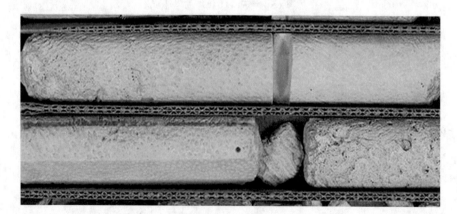

Fig. 5.10 Klinkenberg effect

5.4.2 Permeability, Well Testing Analysis

Permeability can be estimated as a part of pressure build-up analysis. It is calculated using the slope of the straight line in Horner (1951) famous plot (Fig. 5.11) and applying the formula,

$$K_o = 162.6 \, q_o B_o \mu_o / mh$$

where,

K_o = effective permeability to oil, md.

q_o = oil flow rate prior to well shut-in, stb/d.

B_o = oil formation volume factor, bbl/stb.

μ_o = oil viscosity, cp.

m = Horner plot straight line slop, psi/log cycle.

h = net pay thickness of the tested interval, ft.

t_p = production time prior to well shut-in, hours.

Δt = well shut-in time, hours.

It is to be mentioned that the permeability estimated using this approach is the effective permeability to oil. It is the average value for the whole tested interval. The

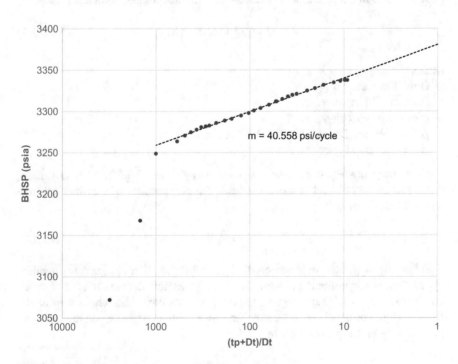

Fig. 5.11 Typical horner plo

effective permeability changes as the fluid saturation changes and this calculated permeability represents the condition at the time of the well testing. Accordingly, this approach is only reliable at the early reservoir life and when the flow around the well is still single phase. These factors should be considered when comparing the well testing calculated permeability with the core measured one.

5.4.3 Permeability, Mathematical Modeling

The main approach to determine permeability is the core measurement. Normally cores do not cover all the reservoir either laterally or vertically. Many researchers suggest different mathematical equations to calculate the permeability from other reservoir rock parameters. Most of these equations based on the assumption that the porous medium consists of a bundle of tubes packed together.

5.4.3.1 Fluid Flow in Cylindrical Tube

About 150 years before Darcy work, Poiseuille (1840) came up with the equation know as Poiseuille's equation to calculate the fluid flow though cylindrical tube;

$$Q = (\pi r^4/8\mu)\,(\Delta P/\Delta L) \tag{5.4.3.1.1}$$

where,
Q = Flow rate, cm^3/s
r = radius of the tube, cm
μ = Fluid Viscosity, poises
ΔP = Pressure Difference ($P_1 - P_2$), dyne/ cm^2
ΔL = Length of the tube, cm.

Comparing this equation with the basic Darcy's law (5.4.3.1.1) and assuming systematic unit system, it is recognized that the permeability of a cylindrical tube can be written as;

$$K = r^2/8 \tag{5.4.3.1.2}$$

Assuming a porous medium consists of a bundle of tubes packed together. Figure 5.12 is a hypothetic porous medium cross section showing bundle of cylindrical. As seen in the same figure, the tubes do not cover the whole area of the cross section. Porosity is introduced to the above formula so that it can represent the permeability of the porous medium;

$$K = \Phi r^2/8 \tag{5.4.3.1.3}$$

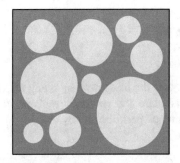

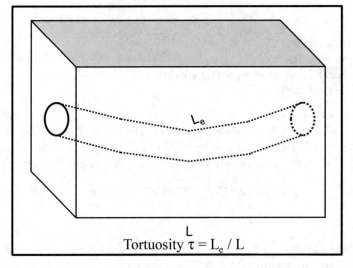

Fig. 5.12 Porous medium consists of a bundle of tubes packed together

where,

K = Permeability, cm^2

r = radius of the tube, cm

Φ = Porosity, fraction.

In the above equation, the tubes of the bundle are assumed to be straight along the porous medium. Obviously, this is not the case. It was suggested to introduce another parameter to take care of this. This parameter is known as the tortuosity (τ) and it takes care about the actual flow path is not straight along the porous medium. **Tortuosity** is defined as the ratio of actual flow path length (L$_e$) to the length (L) of the porous medium in the direction of macroscopic flow (Fig. 5.12). By introducing this dimensionless parameter, Eq. 5.4.3.1.2 will turn to;

$$K = \Phi r^2/8\tau \qquad\qquad (5.4.3.1.4)$$

where,

K = Permeability, cm^2
r = radius of the tube, cm
Φ = Porosity, fraction
τ = Tortuosity, fraction.

The above discussion was the basis used by many researchers to develop mathematical formulae to calculate the reservoir rock permeability. Kozeny (1927) suggested to calculate the rock permeability by,

$$K = \Phi/k_z S_p^2 \qquad (5.4.3.1.5)$$

where,
  K = Permeability
  Φ = Porosity, fraction
  S_p = Internal surface area per unit volume
  k_z = Kozeny constant.

Kozeny equation was then modified by Carman (1937). Table 5.4 shows some famous mathematical formulae to calculate the reservoir rock permeability.

Choo's equation was developed by PETRONAS Carigali in-house to calculates the permeability as a function of porosity and sand, silt and clay content

$$K = 1.65 * 10^7 \Phi_t^{6.243}/10^{(5.5Vcl+2.5Vsilt+1.3Vsand)} \qquad (5.4.3.1.6)$$

where,
  K = Permeability, md
  Φ_t = Total porosity (from open hole logs), fraction
  V_{cl}, V_{silt} and V_{sand} are clay, silt and sand content (fraction) respectively.
  The equation has proved to work properly for Malaysian clastic formation.

5.4.4 Permeability, Correlation

Dislike porosity, the permeability determination is generally limited to cored wells data. The famous Poro-Perm relation is a common practice to make use of the more abundant porosity data to estimate permeability away from well locations. The semi-log poro-perm plot is normally constructed using core analysis data with porosity on linear scale and permeability on the logarithmic scale. A mathematical formula which represents the best fit of the data points can be created. The formula is then used to calculate the permeability all over the reservoir. Generally, a good trend can be obtained for sandstone reservoirs. Due to higher level of heterogeneity, the data points from a carbonate reservoir would be quite scattered (Fig. 5.13) and it is not easy to obtain reliable correlation formula. Figure 5.14, shows the general trends of different lithology data clouds on the poro-perm plot.

Table 5.4 Shows some famous mathematical formulae to calculate the reservoir rock permeability.

Name	Equation	Equation
Solution channel	$K = 0.2 * 10^8 * d^2$	K = permeability (D) d = channel diameter (inch)
Fractures	$K = (0.544 * 10^8 * w^3)/h$	K = permeability (D) h = fracture width (inch) w = fracture aperture (inch)
Timur	$K = (0.136\Phi^{4.4})/(S_{wi})^2$	K = permeability (mD) Φ = porosity (fraction) S_{wi} = irreducible water saturation (%)
Kozeny–Carmen	$K = c\,d^2\,\Phi/(1-\Phi)^2$	K = permeability (mD) Φ = porosity (fraction) c = constant d = median grain size (micron
Berg	$K = 8.4 * 10^{-2} * d^2\,\Phi^{5.1}$	K = permeability (mD) Φ = porosity (fraction) d = median grain size (micron
Van Baaren	$K = 10\,(D_d)^2\,\Phi^{(3.64+m)}\,C^{-3.64}$	K = permeability (mD) Φ = porosity (fraction) D_d = median grain size (micron) C = sorting index m = Archie cementation exponent
RGPZ	$K = (1000\,d^2\,\Phi^{3m})/4\,a\,m^2$	K = permeability (mD) Φ = porosity (fraction) d = weighted geometric mean grain size (micron) m = Archie cementation exponent a = grain packing constant

After Dr. Paul Glover

Fig. 5.13 Poro-permplot for carbonate reservoir

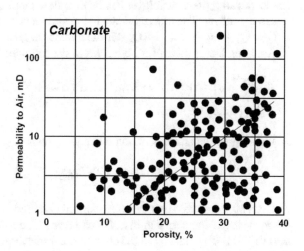

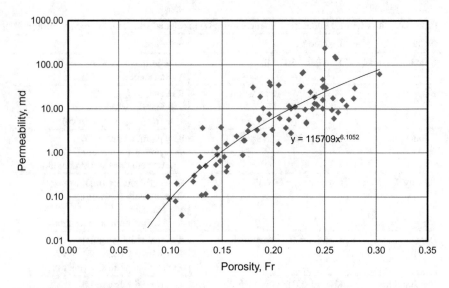

Fig. 5.14 Poro-perm plot

5.5 Permeability Avaraging

As previously highlighted, permeability is a vector quantity that possesses both value and direction. To calculate the average permeability of number of layers, it is essential to consider the direction of flow. Considering the horizontal flow through a number of layers of different permeability and thickness values (Fig. 5.15), the average permeability would be calculated by applying simple flow equation (Darcy's Law). For this **in-parallel** flow, each layer has its own flow rate but all layers share the same pressure gradient. The total flow rate in the system can be calculated by;

$Q = Q_1 + Q_2 + \ldots \ldots + Q_n$ n is the number of layers.

Applying Darcy's Law for both sides of the formula;

$$K_{avg} \sum h/\mu(\Delta P/\Delta L) = K_1 h_1/\mu(\Delta P/\Delta L) + K_2 h_2/\mu(\Delta P/\Delta L) + \ldots \ldots + K_n h_n/\mu(\Delta P/\Delta L)$$

Working up the above equation will end up with

$$K_{avg} = \sum (Kh)/\sum h \tag{5.5.1}$$

It can be recognized that the above formula is simply calculating weighted **arithmetic** average. The permeability of each layer (K_i) has been weighted by its thickness (h_i). It is obvious that in the case of equal thickness layer the equation will be simplified to;

Fig. 5.15 Different thickness LayersIn-parallel flow

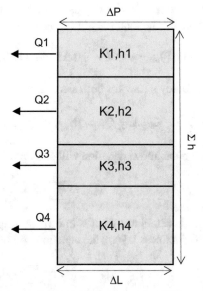

Weighted Arithmetic Averaging $K_{avg} = \Sigma(Kh) / \Sigma h$

$$K_{avg} = \sum(K)/n \qquad\qquad (5.5.2)$$

Let us now consider the vertical flow through the same layers (Fig. 5.16). For this

Fig. 5.16 Different thickness LayersIn-series flow

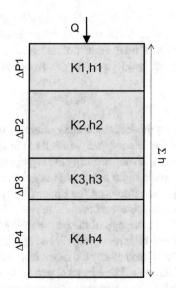

Weighted Harmonic Averaging $K_{avg} = \Sigma h / \Sigma(h/K)$

in-series flow, single flow rate goes through all layers but each layer has its own pressure gradient. The total pressure drop can be calculated by;

$$(\Delta P)_{total} = (\Delta P)_1 + (\Delta P)_2 + (\Delta P)_3 + \ldots\ldots + (\Delta P)_n$$

Applying Darcy's Law for both sides of the formula;

$$Q\mu \sum h/K_{avg}A = Q\mu h_1/K_1 A + Q\mu h_2/K_2 A + \ldots + Q\mu h_n/K_n A$$

Working up the above equation will end up with

$$K_{avg} = \sum h / \sum (h/K) \tag{5.5.3}$$

This formula calculates what is known as weighted **harmonic** average. For the case of equal thickness layers, the equation will be simplified to;

$$K_{avg} = n / \sum (1/K) \tag{5.5.4}$$

Another method of averaging permeability is the **geometric** or **engineering** average. This method is considering that the permeability tends to change in logarithmic or power way and it can be applied for randomly distributed data where no direction of flow needs to be considered.

$$K_{avg} = (w_1 K_1 w_2 K_2 \ldots\ldots w_n K_n)^{1/\sum w} \tag{5.5.5}$$

In this formula, w is the weight that represents the degree of confidence of each single data point. More confident data point is given higher weight than the less confident oned. If all the data points have the same level of confidence, weight of 1.0 is to be applied for all data and the averaging formula will be simplified to;

$$K_{avg} = (K_1 K_2 \ldots\ldots K_n)^{1/n} \tag{5.5.6}$$

Let us now think about some options to upscale. Figure 5.17a shows a group of 36 identical fine blocks that we need to combine (to upscale) into one coarse block. When dealing with permeability averaging, we should think about the flow direction. For the given example the flow is considered to be horizontal flow. In Fig. 5.17b, we initially concentrate on the upper front row of blocks. This row of four blocks exhibits in-series flow and the whole row average permeability should be calculated as the harmonic average of the permeability of the individual blocks. Repeating the procedure for similar rows, we will end up with nine elongated blocks (Fig 5.17c). The permeability of each elongated block has been calculated as the harmonic average of its original individual fine blocks' permeability values. These nine elongated blocks exhibit parallel flow and the average permeability of the whole group should be calculated as the arithmetic average of the individual elongated blocks permeability.

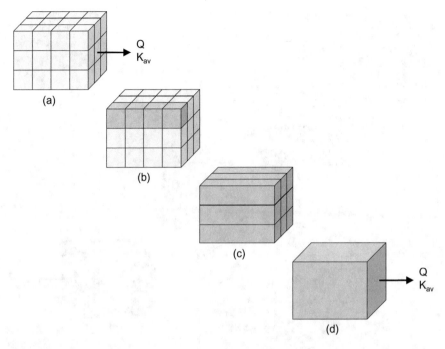

Fig. 5.17 Upscale to reservoir simulation model (Harmonic-Arithmetic)

This procedure of successive averaging is simply defined as **Harmonic-Arithmetic** averaging.

Another option of up scaling is clarified in Fig. 5.18 For this option we start considering the group of nine fine blocks as illustrated in Fig. 5.18b. The blocks in this group exhibit parallel flow and the average permeability for the group should be calculated as the arithmetic average of the permeability of the individual blocks. Repeating the procedure with similar groups, we will end up with four coarser blocks (Fig. 5.18c). The permeability of each coarser block has been calculated as the arithmetic average of its original individual fine blocks' permeability values. These four coarser blocks exhibit in-series flow and the average permeability of the whole group should be calculated as the harmonic average of these individual coarser blocks' permeability. This procedure of successive averaging is simply defined as **Arithmetic-Harmonic** averaging.

The third option to up-scale is to combine the previous two options. After calculating the harmonic-arithmetic (K_{h-a}) and the arithmetic-harmonic (K_{a-h}) averages as explained, the geometric average of the two values can be calculated as the final estimated average permeability of the whole system.

$$K_{avg} = (K_{h-a}K_{a-h})^{0.5}$$

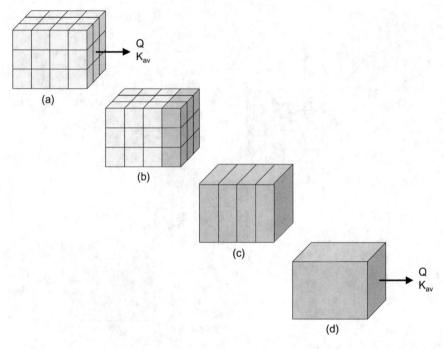

Fig. 5.18 Upscale to reservoir simulation model (Arithmetic-Hamonic)

The above methods are just explained to show the importance and effect of considering the flow direction in calculating the average permeability. The available reservoir modeling software would provide more up-scaling approaches. However, the idea of considering fluid flow direction is always the driving factor.

5.6 Permeability for Reservoir Simulator

Permeability is introduced to reservoir simulator as values in milli-darcy in array form. It is normally imported from the output of the geological model. The simulator requires three different groups of permeability values, two groups for horizontal permeability (PERMX and PERMY) and one group for the vertical permeability (PERMZ). By giving such information, the simulator will be provided by three orthogonal permeability values for each model block. Each provided permeability value must be equal to or greater than zero. It is not recommended to provide zero permeability to any reservoir model block. Zero permeability blocks are really inactive ones. However, the simulator will treat them as normal blocks and consume unnecessary computer time to perform normal flow calculation for them. The best practice to treat inactive blocks in the reservoir model is to set their porosity or pore volume to zero (see Sect. 3.5). Here in is an example of porosity introduced to simulator for a hundred blocks one dimensional reservoir model.

PERMX

> 30.0 30.0 30.0 30.0 30.0 30.0 30.0 30.0 30.0 30.0
> 30.0 30.0 30.0 30.0 30.0 30.0 30.0 30.0 30.0 30.0
> 55.0 55.0 55.0 55.0 55.0 55.0 55.0 55.0 55.0 55.0
> 55.0 55.0 55.0 55.0 55.0 55.0 55.0 55.0 55.0 55.0
> 55.0 55.0 55.0 55.0 55.0 55.0 55.0 55.0 55.0 55.0
> 80.0 80.0 80.0 80.0 80.0 80.0 80.0 80.0 80.0 80.0
> 80.0 80.0 80.0 80.0 80.0 80.0 80.0 80.0 80.0 80.0
> 80.0 80.0 80.0 80.0 80.0 80.0 80.0 80.0 80.0 80.0
> 100. 100. 100. 100. 100. 100. 100. 100. 100. 100.
> 100. 100. 100. 100. 100. 100. 100. 100. 100. 100.

PERMY

> 20 ∗ 30.0 30 ∗ 55.0 30 ∗ 80.0 20 ∗ 100.0

PERMZ

> 20 ∗ 3.0 30 ∗ 5.5 30 ∗ 80.0 20 ∗ 10.0

Exercises:

1. Drive the formulae for calculating horizontal and vertical weighted average permeability for a number of layers.
2. An oil reservoir consists of four layers with properties as shown in the table:

Layer number	1	2	3	4
Thickness (ft)	3	5	2	7
Permeability (md)	100	70	20	80

For each layer, the horizontal and vertical permeabilities are the equal.

Calculate the average horizontal permeability and the average vertical permeability for this reservoir.

Give comment on results

3. The permeability of a core plug was measured in laboratory as 58 md using air. If the mean pressure applied during experiment was 25 psi, calculate the absolute permeability of the plug.

References

Carman, P.C. (1937). Fluidflow through granular beds. *Transactions of the Institution of Chemical Engineers, 15*, 150.

Darcy, H. (1856). Les fontaines publiques de la ville de Dijon. Cf. M. Muskat *The flow of homogeneous fluids through porous media* (New York 1937).

Heid, J.G., McMahon, J.J., Nielsen, R.F., & Yuster, S.T. (1950). *Study of the permeability of rocks to homogeneous fluids.* Paper presented at the Spring Meeting of the Southwestern District, Division of Production, Dallas, Texas (March 1950); in API Drilling and Production Practice, 230–246.

Horne, D. R. (1951). *Pressure build-up in wells.* World Petroleum Congress.

Jones, F.O., Owens, W.W. (1979). *A laboratory study of low permeability gas sands.* Paper SPE 7551 presented at the 1979 SPE Symposium on Low-Permeability Gas Reservoirs, May 20–22, 1979, Denver, Colorado.

Klinkenberg, L. J. (1941). *The permeability of porous media to liquids and gases.* New York: Drilling and Production Practice, API-41-200.

Kozeny, C. (1927). Uber die kapillare Leitung des Wassers im Boden (Aufstieg Versickerung und Anwendung auf die Bewasserung. Sitz. Ber. Akad. Wiss, Wien, Math. Nat. (Abt. Iia), 136a (1927), p. 271.

Muskat M. (1937). *The flow of homogeneous fluids through porous media* (p. 69). New York and London: McGraw-Hill.

Poiseuille, J. (1840). Recherches expérimentales sur le mouvement des liquides dans les tubes de très petits diamètres [Experimental research on the movement of liquids in capillary of very small diameters]. *Comptes rendus hebdomadaires des séances de l'Académie des sciences, T11*, 1041–1048.

Sampath, K., Keighin, C.W. (1981) *Factors affecting gas slippage in tight sandstones.* Paper SPE 9872 presented at the 1981 SPE/DOE Low Permeability Symposium, May 27–29, 1981, Denver, Colorado.

Tarek, G. (2019). *Fundamentals of reservoir rock properties.* Springer International Publishing. ISBN 978-3-030-28139-7.

Wu, Y. -S., & Pruess, K. (1998). *Gas flow in porous medium with Klinkenberg effects.* Berkeley, CA 94720, USA: Earth Science Division, Lawrence Berkeley National Laboratory.

Chapter 6
Rock Compressibility

6.1 Rock Compressibility Definition and Types

Compressibility of any material is defined as the relative change of volume caused by unit change of applied pressure or stress. The units used for compressibility is the reciprocal of pressure unit. In practical field applications, the unit used is 1/psi which can also be expressed as psi^{-1} or sip. Compressibility is expressed mathematically as,

$$C = -1/V(\Delta V/\Delta P) \tag{6.1.1}$$

where,

C = material compressibility, psi^{-1}
V = initial material volume
ΔP = pressure change, psi
ΔV = volume change caused by pressure change ΔP.

The negative sign which appears at the right-hand side of the above formula indicates that the volume decreases as the effective pressure increases. By effective pressure, it is meant the stress or pressure which directly affecting the material volume. It is common to remove this sign for simplicity. However, it is important to keep understanding that the volume decreases as the effective pressure increases. So, the common form of the equation defining compressibility is,

$$C = 1/V(\Delta V/\Delta P) \tag{6.1.2}$$

It is also expressed in differential formula as,

$$C = 1/V(\partial V/\partial P) \tag{6.1.3}$$

A. M. Badawy and T. A. A. O. Ganat, *Rock Properties and Reservoir Engineering: A Practical View*, Petroleum Engineering,
https://doi.org/10.1007/978-3-030-87462-9_6

For reservoir porous rock, there are three different types of compressibility. The **Rock Bulk Compressibility** (C_B) is the relative bulk volume change caused by a unit change of pressure. The **Rock Matrix (Grains) Compressibility** (C_r) is defined as the relative matrix (grains) volume change caused by a unit change of pressure. Similarly, the **Pore Space Compressibility** (C_p) is the relative pore volume change caused by a unit change of pressure. The rock bulk compressibility and the rock matrix compressibility are normally very small compared to the pore space compressibility. Consequently, the **Effective Formation Rock Compressibility** (C_f) is considered as equal to the pore space compressibility (C_p).

$$C_f = 1/V_p(\Delta V_p/\Delta P) \tag{6.1.4}$$

where,

C_f = effective rock compressibility.
V_p = rock pore volume.
ΔV_p = rock pore volume change.
ΔP = pressure chang.

In this text, the rock compressibility, effective rock compressibility and por volume compressibility are considered to be the same. Normally, reservoir rock compressibility ranges from 3.0×10^{-6} to 25×10^{-6} psi^{-1}. However, values of up to 60×10^{-6} psi^{-1} or even more have been recorded.

Rock compressibility is an important parameter different oil and gas industry activities. It is an important parameter in designing well drilling programs. It is crucial parameter in completion design to avoid and overcome sand production problems. Besides, it's very significant for modelling the fluids flow in reservoir, and predicting the future oil and gas production.

6.2 Stresses (Pressures) Affecting Reservoir Rock

To understand the effect of formation rock compressibility on reservoir behavior, we need to simply understand the stresses (pressures) affecting the reservoir rock. For a hydrocarbon reservoir buried thousands of feet deep, the reservoir rock grains are subject to external stress caused by the weight of the overlaying formations. This stress (pressure) is known as the **overburden pressure** or lithostatic pressure (P_o). The typical gradient of the overburden pressure (lithofacies gradient) is 1.0 psi/ft.

P_o = pressure caused by the weight of overlaying formation
$P_o = 1.0 \times \text{depth}$
$\tag{6.2.1}$

where,

P_o = overburden pressure, psi.

depth = vertical depth to reservoir, ft.

Equation 6.2.1 is generally applicable for onshore reservoirs. For offshore reservoirs, especially for deep water ones, the equation should be modified to take care of the weight of the sea water.

$$P_o = \text{pressure caused by the weight of water} +$$
$$\text{pressure caused by the weight of overlaying formation}$$
$$P_o = \text{sea water gradient x water depth} +$$
$$1.0 \times (\text{ depth - water depth }) \qquad (6.2.2)$$

On the other hand, the reservoir rock matrix is subject to internal pore space pressure which is normally referred as the **reservoir pressure** (P_r). If not directly measured, this pressure is estimated using the area hydrostatic gradient. A typical value for hydrostatic gradient is 0.433 psi/ft.

$$P_r = 1.0 \times \text{depth} \qquad (6.2.3)$$

where,
P_r = reservoir, psi
depth = vertical depth, ft.

The difference between these two stresses (pressures) is the effective overburden pressure (P_{oe}) and it is the net stress that affects the hydrocarbon reservoir rock (Fig. 6.1)

$$P_{oo} = P_0 - P_r \qquad (6.2.4)$$

This equation assumes that the external stress that affects the reservoir rock consists only of the vertical overburden pressure (P_o or σ_v). Actually, the reservoir rock is also affected by lateral stresses, σ_x and σ_y (Fig. 6.2). Reasonable approximation of these lateral stresses is given by,

$$\sigma_x = \sigma_y = 2/3\sigma_v \qquad (6.2.5)$$

The average external stress is calculated by,

$$\sigma_n = (\sigma_v + \sigma_x + \sigma_y)/3 = 7/9 * (\sigma_v) = 7/9^*(P_o) \qquad (6.2.6)$$

Fig. 6.1 Overburden &
reservor pressures

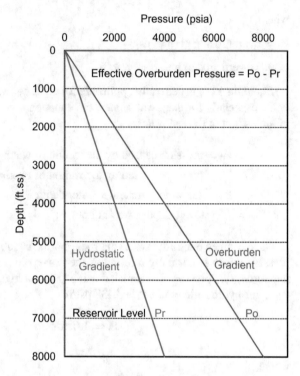

Fig. 6.2 Vertical and
horizontal external stresses

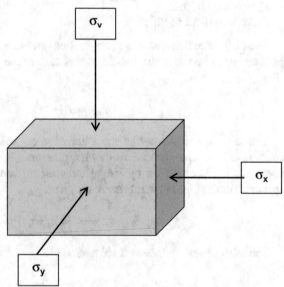

where,

σ_x and σ_y are the external lateral stress components, psi.

σ_v is re the external vertical stress components, psi.

σ_n is the average external lateral stress components, psi.

Subsequently, more accurate effective overburden stress (pressure) is given by,

$$P_{oo} = 7/9(P_o) - P_r \tag{6.2.7}$$

6.3 Effect of Reservoir Depletion on Pore Volume

Basically, the reservoir engineer is interested in how much the reservoir change would affect the pore volume or the porosity and how much this would affect the hydrocarbon recovery. During reservoir depletion, the reservoir pressure (internal stress) decreases while the overburden pressure (external stress) remains almost constant. Subsequently, the effective overburden pressure increases. Practically the increase of the net overburden pressure is equal to the decrease of reservoir pressure. It is possible to notice two effects that affect the pore volume due to depletion. The first one is the expansion of the rock matrix due to reservoir pressure decrease. The second one is the reduction of the bulk volume due to the increase of the effective overburden pressure. It is clear that these two effects will cause reduction in the pore volume (Fig. 6.3). The current pore volume can be estimated by referring to Eq. 6.1.4,

$$C_f = 1/V_p(\Delta V_p/\Delta P)$$

This equation can be worked up to calculate the current pore volume,

$$V_p = V_{po}[1 + C_f(P - P_o)] \tag{6.3.1}$$

The bulk volume change is very small as compared to the pore volume change and can be ignored. Subsequently, the pore volume terms in Eq. 6.1.4 can be replaced by porosity terms,

$$C_f = 1/\Phi(\Delta\Phi/\Delta P) \tag{6.3.2}$$

Then, the current porosity can be calculated by,

$$\Phi = \Phi_0[1 + C_f(P - P_o)] \tag{6.3.3}$$

where,

C_f Formation Compressibility

P_o Original Reservoir Pressure

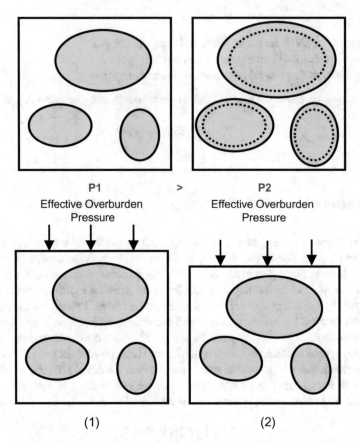

Fig. 6.3 Effect of reservoir pressure drop on pore volume

V_{po} Original Pore Volume
V_p Pore Volume at current pressure P
ϕ_o Original Porosity
ϕ Porosity at current pressure P.

For compressibility range of 3×10^{-6} to 25×10^{-6} psi^{-1}, a reservoir pressure drop of 1000 psi will cause 0.3–2.5% pore volume reduction (Eq. 6.3.1). This looks to be quite small effect. However, this should not be taken as a reason to ignore the impact of rock compressibility on reservoir performance. Ignoring pore volume compressibility in material balance calculation would result in over estimation of oil in place and calculation of excessive volume of water influx.

To understand the effect of rock compressibility on predicting reservoir performance, let us consider the simple one-dimension reservoir model shown in Fig. 6.4 This simple model is built of 101 similar blocks. The model is designed to simulate a homogeneous reservoir of 100 md permeability and 15% porosity. The gas oil contact (GOC) and oil water contact (OWC) were set at shallow and deep levels respectively

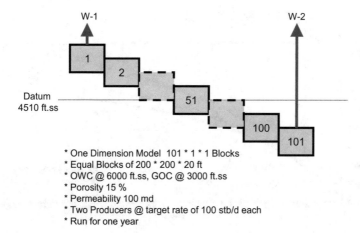

Fig. 6.4 One-dimension test model

so that the reservoir is not in contact with any gas cap or water aquifer. Two wells were placed at the far ends of the reservoir with 100 stb/day target rate each. The model was run to predict one year performance using rock compressibility values of 3.0×10^{-6}, 5.0×10^{-6} and 10.0×10^{-6} psi^{-1} respectively. Figures 6.5, 6.6 and 6.7 show the effect of different rock compressibility values on reservoir pressure, oil production rate and cumulative oil production. It can be noticed that as the effective rock compressibility increases, the reservoir pressure reduction decreases and the oil recovery increase. Simply, it can be concluded that the effective rock compressibility acts as a reservoir drive mechanism. This mechanism is more significant while the

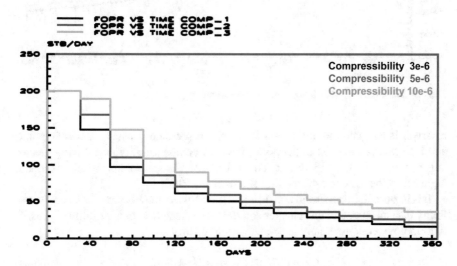

Fig. 6.5 Sensitivity to rock compressibility oil production rate

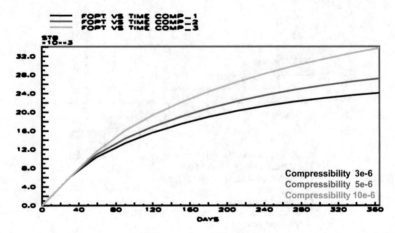

Fig. 6.6 Sensitivity to rock compressibility cumulative oil production

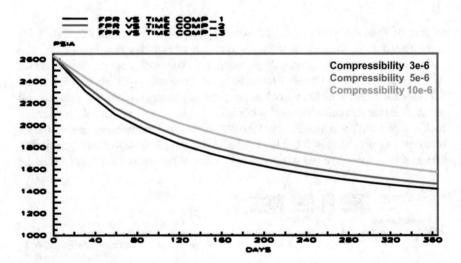

Fig. 6.7 Sensitivity to rock average reservoir pressure

reservoir is still undersaturated. As the reservoir pressure reaches the bubble point and free gas start to exist in the pores, the rock compressibility as a driving mechanism starts to be less significant. This is because, the gas compressibility is much higher than the rock compressibility (Frick et al., 1962).

It is important to be careful and differentiate between the effective rock compressibility (C_f) discussed in this chapter and the total effective compressibility (C_t) used in fluid flow work especially in pressure build up analysis.

$$C_t = C_f + S_w C_W + S_o C_o + S_g C_g \qquad (6.3.4)$$

where,

C_t = total effective compressibility
C_f = effective rock compressibility
C_w = water compressibility
C_o = oil compressibility
C_g = gas compressibility
S_w = water saturation, fraction
S_o = oil saturation, fraction
S_g = gas saturation, fraction.

6.4 Rock Compressibility Determination

Two approaches are available to determine the effective rock compressibility. These are the laboratory core measurement and the available correlations. Rock compressibility is not considered in normal core analysis programs. This could be due to high cost and time consumption. Many reservoir studies rely on determining rock compressibility using available correlations. However, it is thought that this attitude should be changed. The practical experience shows that correlation can be unreliable in many cases especially where handling unconsolidated formations.

6.4.1 Rock Compressibility Determination, Laboratory Experiments

The effective formation compressibility is measured in laboratory as a part of special core analysis (SCAL) work program. Carpenter and Spencer (1940) are the earliest reported researchers who attempted to measure reservoir rock compressibility. They conducted compressibility measurement on consolidated sandstone samples from East Texas.

To determine the effective rock compressibility, the laboratory experiment is designed to simulate the effect of the net overburden stress on reservoir rock. The rock compressibility determination is based on measuring the variation in rock pore volume of core sample due to variation in effective over burden pressure. The experiment starts by measuring the pore volume of the sample at atmospheric pressure and reservoir temperature. The core sample is then placed in soft cupper or rubber sleeve in a core holder and saturated with brine. Figure 6.8 is a schematic presentation of the experiment. The external sleeve pressure simulates the overburden pressure (P_o) and internal pore pressure simulates the reservoir pressure (P_r). The two pressures are increased in alternative increments to achieve stabilization. Then, the internal pressure is reduced to simulate the reservoir depletion and the corresponding reduced pore volume is recorded (refer to previous Sect. 6.3). The pore volume reduction is

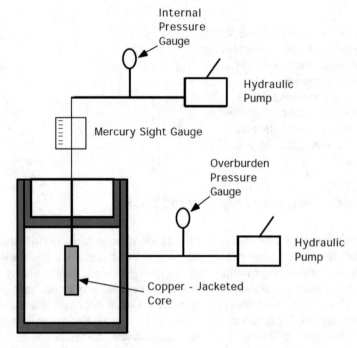

Fig. 6.8 Rock compressibility measurement

equal to the volume of the expelled brine. Rock compressibility values are calculated using a plot of pore volume versus effective overburden pressure. The slope of the curve (Fig. 6.9) represents the change of pore volume per unit change of effective overburden pressure ($\partial V/\partial P$) and the compressibility can be calculated at any specific

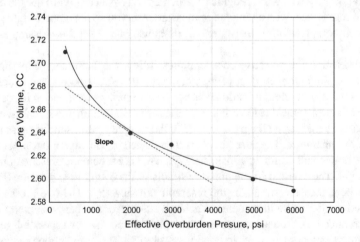

Fig. 6.9 Pore volume versus effective overburden pressure

overburden pressure. As a completion to the procedure, the resultant compressibility is corrected to uniaxial loading, by multiplying by factor 0.61, according to Teeuw (1971).

6.4.2 Rock Compressibility Determination, Correlation

Researchers have been working establish correlations to estimate effective reservoir rock compressibility as a function of porosity. The normal trend of such correlations showed that compressibility decreases as porosity increases. This trend was recognized for consolidated sandstone and limestone samples.

Hall (1953) established a graphical correlation between effective rock compressibility (C_f) and porosity (ϕ) using twelve limestone and consolidated sandstone samples from different reservoirs. This correlation (Fig. 6.10) shows very good fitting for the used samples and it follows the expected trend of the compressibility-porosity relationship. The attractive thing about this correlation is that it is applicable for both sandstone and limestone. This correlation was used by reservoir engineers for long time and many oil companies did not appreciate the necessity of measuring formation compressibility in laboratory. Hall's graphical relation would suggest the correlation formula,

$$C_f = \left(1/\Phi^{1.782}\right) \tag{6.4.2.1}$$

where,
C_f = effective rock compressibility, 10^{-6} psi^{-1}
ϕ = porosity, fraction.

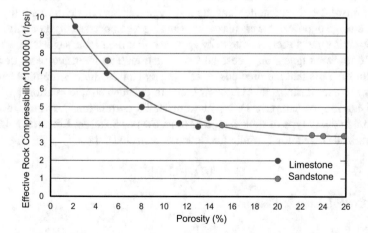

Fig. 6.10 Effective rock compressibility versus porosity (after hall)

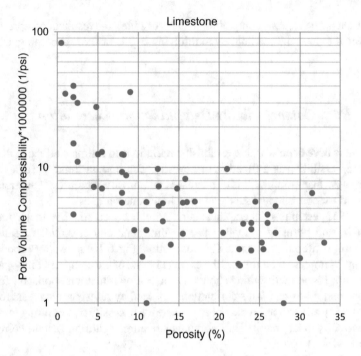

Fig. 6.11 Effevtive rock compressibility versus porosity (after Newman)

Newman (1973) used numerous samples of different types (limestone, consolidated sandstone, unconsolidated sandstone and friable sandstone) to establish more detailed correlation. Figures 6.11, 6.12, 6.13 and 6.14 are graphical presentation of the data used by Newman. From these graphs, it can be recognized the possibility of getting reasonable correlations for limestone, consolidated and unconsolidated sandstones. The unconsolidated sandstone data suggest large increase of compressibility with limited change of porosity. This makes considering a compressibility-porosity correlation for this type of rock is quite risky approach. For the friable sandstone, the data scatter does not suggest any reasonable correlation. The intensive work done by Newman suggests that the previous correlation by Hall is highly simplified and can be tricky to use. It seems that Hall was lucky to get such simple relationship using limited number of samples. Different from Hall's correlation, Newman research showed that every type of rock should have its own correlation. Equation 6.4.2.2 is a generalized correlation formula for Newman collected data,

$$C_f = a/[1 + cb\Phi]^{(1/b)} \qquad\qquad (6.4.2.2)$$

For Consolidated Sandstone

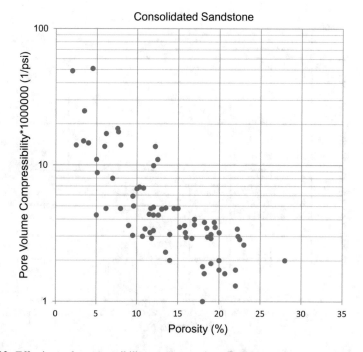

Fig. 6.12 Effective rock compressibility versus porosity (after Newman)

$$a = 97.31 \times 10^{-6}$$
$$b = 0.69993$$
$$c = 79.8181$$

For Limestone

$$a = 0.8535$$
$$b = 1.075$$
$$c = 2.202 \times 10^{6}$$

where,
C_f = effective rock compressibility, psi^{-1}
ϕ = porosity, fraction.

Horne (1990) used the same data collected by Newman to establish exponential formulae to correlate the pore volume compressibility with porosity,

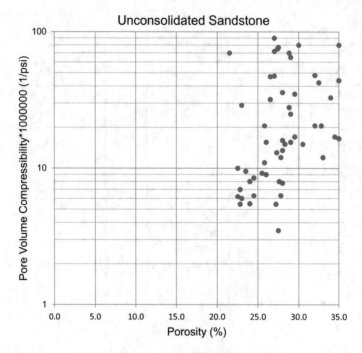

Fig. 6.13 Effective rock compressibility versus porosity (after Newman)

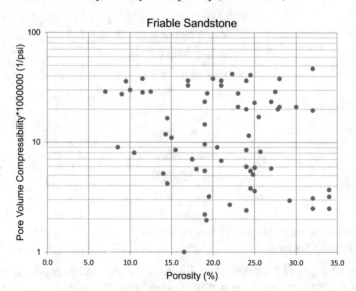

Fig. 6.14 Effective rock compressibility versus porosity (after Newman)

For consolidated sandstone,

$$C_f = \exp\left(5.118 - 36.26\Phi + 63.98\Phi^2\right) \qquad (6.4.2.3)$$

For limestone,

$$C_f = \exp\left(4.026 - 23.07\Phi + 44.28\Phi^2\right) \qquad (6.4.2.4)$$

where,
C_f = effective rock compressibility, 10^{-6} psi^{-1}
ϕ = porosity, fraction.

Figures 6.15 and 6.16 show comparison between Hall and Neman correlations for consolidated sandstone and limestone.

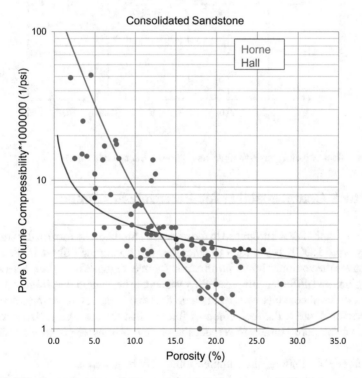

Fig. 6.15 Effective rock compressibility versus porosity (home versus hall)

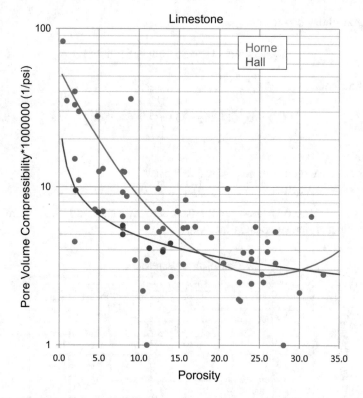

Fig. 6.16 Effective rock compressibility versus porosity (home versus hall)

6.5   Rock Compressibility for Reservoir Simulator

Rock compressibility is introduced to the reservoir simulator in a form of data records. The key word ROCK is the used to enter the rock compressibility data. Each data record gives the compressibility information for one region. The regions here are the PVT regions and different compressibility values can be entered for different regions. Each data record contains two parameters, first one is the reference pressure (P_{ref}) and the second one is the rock compressibility value for the region. The simulator will adjust the pore volume to respect the change of pressure from the reference pressure.

For three PVT regions, the compressibility will be given as,

ROCK.
    4500.0 3.5E-06/
    4500.0 4.0E-06/
    4000.0 1.0E-06/

Exercise:

1. Rock compressibility experiment was achieved on a core plug from a newly discovered reservoir. The given table shows the recorded data.

Effective overburden pressure (psi)	Sample pore volume (cubic centimeter)
400	4.02
1000	3.99
2000	3.94
3000	3.91
4000	3.88
5000	3.86
6000	3.84

    Calculate the effective rock compressibility at effective overburden pressures of 1000, 3000 and 5000 psi. Comment on results.

2. An oil reservoir was characterized as consolidated sandstone reservoir. The
3. original average porosity of the reservoir rock was estimated as 20%, the original reservoir pressure is 5000 psia and the current reservoir pressure is 4500 psia.

    Calculate the reservoir formation compressibility and current average porosity (Use Hall and Horne Correlations).

References

Carpenter, C. R., & Spencer, G. R. (1940). *Measurements of compressibility of consolidated oil-bearing sandstones*. Bureau of Mines RI 3540.

Frick, T. C., & Taylor, R. W. (1962). Petroleum production handbook. In *Reservoir engineering* (vol. II). Dallas, Texas: Society of Petroleum Engineers of AIME.

Hall, H. N. (1953). Compressibility of reservoir rocks. *Journal of Petroleum Technology., 5*(1), 17–19.

Horne, R. N. (1990) *Modern well test analysis, a computer-aided approach*. Petroway Inc.

Newman, G. H. (1973). Pore-volume compressibility of consolidated, friable, and unconsolidated reservoir rocks under hydrostatic loading. *Journal of Petroleum Technology., 25*(2), 129–134.

Teeuw, (1971). Prediction of formation compaction from laboratory compressibility data (SPH paper 2973). *Society of Petroleum Engineers Journal, 11*, 263–271. https://doi.org/10.2118/297 3-PA.

Chapter 7
Wettability

7.1 Wettability

Wettability is defined as the preference of one fluid to spread or adhere to a solid surface in the existence of other immiscible fluids. In two phase system, the phase which tends to adhere to the rock (solid) surface is known as the wetting phase. The other phase is the non-wetting phase. The wetting phase tends to occupy the smaller pores and subsequently more difficult to move as compared to the non-wetting phase. It is believed that reservoir rocks are originally water wet. This sounds logic because they are originally fully saturated with water. After oil migration to the reservoir, the reservoir rock wettability may change along the time towards oil wet. This change is due to the solid rock surface direct contact with some types of crudes. Most reservoir rocks are of intermediate wettability. However, sandstone reservoirs tends to be more water wet while carbonate reservoirs tend to be more oil wet. Gas is understood to be non-wetting phase in the existence of oil and/or water. Subsequently, for gas reservoirs, the reservoir rock is to be considered water wet (Tarek Tarek 2019).

Examining equations which control the fluid flow in porous media, no room for rock wettability can be noticed. However, through its effect on other combined rock fluid properties (capillary pressure and relative permeability), wettability has great influence on the flow of oil, gas and water through the porous media and subsequently to hydrocarbon recovery (Brooks and Corey, 1966).

In the discussion about wettability, we should consider the relation with **contact angle**. When two immiscible fluids come in contact with a solid surface, their interface will make an angle with that surface. This angle is known as the contact (wetting or dihedral) angle. The contact angle value relies on the relative wettability of the solid surface (rock) to either one of the two fluids (Fig. 7.1). Actually, laboratory measurement of this contact angle represents the essential method of determining the rock wettability (Anderson, 1986).

Different methods have been conducted to estimate relative rock wettability through laboratory experiments. Amott and USBM (United States Bureau of Mines)

© The Author(s), under exclusive license to Springer Nature Switzerland AG 2022   75
A. M. Badawy and T. A. A. O. Ganat, *Rock Properties and Reservoir Engineering: A Practical View*, Petroleum Engineering,
https://doi.org/10.1007/978-3-030-87462-9_7

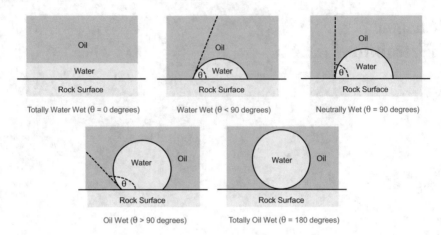

Fig. 7.1 Wettability and contact angle

could be the most common laboratory methods for estimating the rock wettability (Amott 1959). However, and due its complexity, there is no comprehensive technique that can offer accurate estimation of the reservoir rock wettability especially in the neutral range (intermediate wettability). Table 7.1 shows comparison of wettability index as expected from different laboratory methods.

In addition to laboratory methods, there are some rules to estimate the reservoir rock relative wettability from other properties measurements. It is a common practice to get some understanding of the relative reservoir rock wettability by examining the water–oil relative permeability curves. Table 7.2 shows guidelines to estimate the

Table 7.1 After Paul Glover

	Oil-wet	Intermediate	Water-wet
Amott wettability index water ratio	0	0	>0
Amott wettability index oil ratio	> 0	0	0
Amott-Harvey wettability index	−1.0 to −0.3	−0.3 to 0.3	0.3 to 1.0
USBM wettability index	About −1.0	About 0	About 1.0
Minimum contact angle (°)	105–120	60–75	0
Maximum contact angle (°)	180	105–120	60–75

Table 7.2 xxx

	Oil-wet	Intermediate	Water-wet
S_w @ $K_{rw} = K_{row}$	<50%	~50%	>50%
S_{wirr}	<15%	15–25%	>25%
$K_{rw\,max}/K_{ro\,max}$	>0.5	0.3–0.5	<0.3

relative wettability of the reservoir rock applying this practice. These guidelines are based on the rule of thumb suggested by Craig (1993).

As mentioned above, the estimation of wettability is quite non-conclusive. From practical point of view, it seems not very crucial to include wettability measurement in normal SCAL program. This is understood remembering that there is no room for wettability in any fluid flow equation. Its effect on fluid flow and subsequently on estimating recovery goes through its effect on other properties like relative permeability and capillary pressure. However, considering wettability measurement in the SCAL program can be of special importance where enhanced oil recovery (EOR) techniques are expected in the reservoir development plan.

References

Amott, E. (1959). Observations relating to the wettability of porous rock. *Trans. AIMB, 216*, 156-162.

Anderson, W. G. (1986, November). Wettability literature survey-Part 2: wettability measurement. SPE 13933, *Journal of Petroleum Technology*.

Brooks, R. H., & Corey, A. T. (1966). Properties of porous media affecting fluid flow. *Journal of Irrigation and Drainage Division, Proceeding of American Society of Civil Engineers (ASCE)*, 61–88.

Craig, Forest F., Jr. (1993, January). *The reservoir engineering aspects of water flooding*. Monograph Volume 3 of the Henry L. Dorothy Series. Society of Petroleum Engineers of AIME.

Tarek, G. (2020). *Technical guidance for petroleum exploration and production plans*. Springer International Publishing. ISBN 978-3-030-45250-6.

Chapter 8
Capillary Pressure

8.1 Forces Affecting Two Cotiguous Immisible Fluids

To understand the phenomenon of the capillary pressure, let us first think about the forces affecting two contiguous immiscible fluids. Naturally, when two immiscible fluids are placed in one container, the denser fluid will completely lie below the less dense one. This is what simply understood as buoyancy or gravitational forces. However, if the same two fluids are placed in the pore space of a porous medium, their distribution will not rely only on their densities. There will be a zone where the two fluids will co-exist in the same pores in appreciable amounts. This behavior is caused by the interfacial action between the two fluids and the solid surface of the capillary system. This is what described as the capillary forces.

Interfacial tension is caused by the attraction between the fluid's molecules by various intermolecular forces. Let us consider the surface separating air and water (Fig. 8.1) and try to understand the forces affecting the water molecules. Each molecule inside the bulk of the fluid (water) is pulled equally in all directions by other molecules, resulting in a net force of zero. A molecule at water surface is pulled inwards by the similar molecules deeper inside the bulk of the fluid and is not attracted as intensely by the ones of the neighboring fluid (air). Therefore, the surface molecules are subject to an inward force which is balanced only by the fluid's resistance to compression. However, there is a driving force to diminish the surface area, and in this respect the fluid surface acts as a stretched elastic membrane and the fluid squeezes itself together to reach the locally lowest surface area possible. The water strider (Fig. 8.2) with its tiny long feet can walk on the water surface without disturbing that elastic membrane. Some people are skill enough to put a steel needle on the water surface without disturbing that membrane. The phenomenon of interfacial tension is alternatively known as **surface tension** if one of the two fluids is gas.

© The Author(s), under exclusive license to Springer Nature Switzerland AG 2022
A. M. Badawy and T. A. A. O. Ganat, *Rock Properties and Reservoir Engineering: A Practical View*, Petroleum Engineering,
https://doi.org/10.1007/978-3-030-87462-9_8

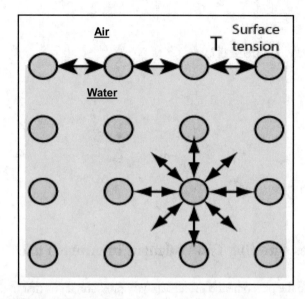

Fig. 8.1 Forces affecting water molecules

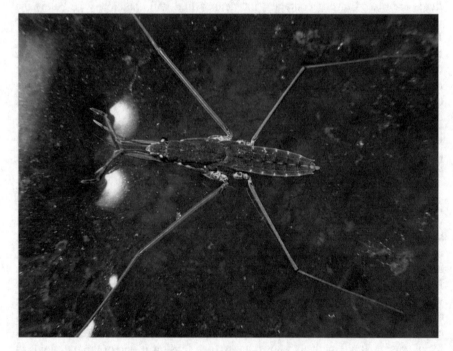

Fig. 8.2 Water strider

8.2 Capillary Pressure Definition and Concept

When two immiscible fluids co-exist in a system of capillaries, the combined effect of the interfacial tension between the two fluids and the curvature due to capillaries cause each fluid to have different pressure. The difference between the two fluids' pressures is defined as the **Capillary Pressure**. The non-wetting phase (fluid) pressure is greater than the wetting phase (fluid) pressure. It was found that the capillary pressure changes with the relative saturation of the two fluids. This observed relation between capillary pressure and fluid saturation establishes the basis for setting the initial fluid saturation in any hydrocarbon reservoir (Tarek, 2019).

Let us consider immersing a capillary tube in a container filled with water (Fig. 8.3). The water (wetting phase) will rise (imbibe) in the capillary tube replacing the air (non-wetting phase) to some level due to the capillary forces. As discussed previously (Chap. 7), the interface between the two phases will make a contact angle with the solid surface of the capillary tube. The capillary forces acting on the peripheral of the capillary tube can be calculated using,

$$F = 2\pi r\sigma \cos\theta \tag{8.2.1}$$

Then the capillary pressure can be calculated by,

$$P_c = 2\pi r\sigma \cos\theta/\pi r^2$$

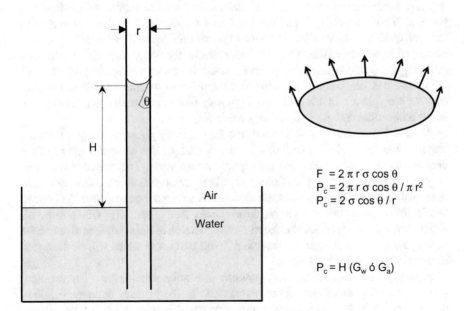

$$F = 2\pi r\sigma \cos\theta$$
$$P_c = 2\pi r\sigma \cos\theta / \pi r^2$$
$$P_c = 2\sigma \cos\theta / r$$

$$P_c = H (G_w \, ó \, G_a)$$

Fig. 8.3 Capillary pressure calculation

$$P_c = 2\sigma \cos \theta / r \qquad\qquad (8.2.2)$$

where,

P^c capillary pressure, dyne/cm^2
σ interfacial tension, dyne/cm
r capillary tube radius, cm.

It can be recognized that the water column inside the capillary tube is balanced under two different forces. The capillary force (up wards) acts to keep the water inside the tube while the gravity or buoyancy force (down wards) tries to push it back into the container. Understanding this balance, the capillary pressure can be calculated by,

$$P_c = H(G_w - G_a) \qquad\qquad (8.2.3)$$

where,

H = height of the water column
G_w = water gradient
G_a = air gradient.

From the two Eqs. (8.2.2) and (8.2.3), it is concluded that the capillary pressure increases and subsequently the water rise in the capillary tube increases as the capillary tube radius decreases (Fig. 8.4). Understanding that the capillary tube simulates pores of porous medium, it can be stated that the capillary pressure increases as the rock quality decreases. In the above explanation (Fig. 8.3), the capillary tube is immersed in water (wetting phase). Consequently, the water rises up into the tube replacing air (non-wetting phase). This process is known as **capillary attraction**. Let us now consider the capillary tube is immersed in a container filled with mercury (non-wetting phase). In this case the air (wetting phase) will push down the mercury in the process defined as **capillary repulsion** (Fig. 8.5).

As mentioned above, it was found that the capillary pressure changes with the relative saturations of the two fluids. From Eq. (8.2.3), it is clear that the capillary pressure can be easily converted to height and vice versa. The most common and maybe the most useful application of capillary pressure is to combine these two relationships to distribute initial saturation in any hydrocarbon reservoir. As known, this is an essential task in reservoir development study especially where reservoir simulation is the study tool. Other important application of this relation is to define the pore size distribution within the reservoir. These two applications will be discussed in more details in coming sections.

Figure 8.6 is a typical capillary pressure saturation relationship. The capillary pressure curve in this figure is the one defined as the primary drainage capillary pressure curve. Other capillary pressure curves will be discussed in later section. The primary drainage capillary pressure curve simulates the process of oil migration to water filled formation and it is the one used to distribute initial saturation in

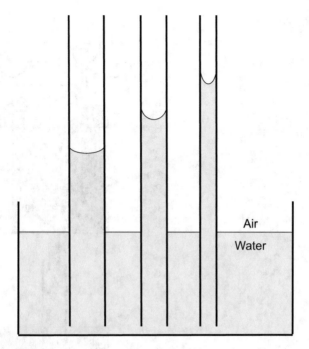

Fig. 8.4 Capillary pressure for different size capillary tubes

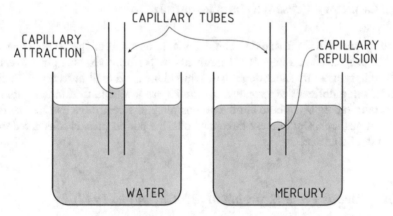

Fig. 8.5 Capillary attraction & capillary repulsion

the reservoir. Two special points on the curve that should be explained here. The irreducible water saturation (S_{wirr}) which is the minimum possible water saturation in the reservoir. The threshold (P_{th}), displacement or entry pressure is the pressure needed for the oil phase to enter and displace water from the largest pore throat pores.

Fig. 8.6 Capillary pressure
determination porous plate
experiment

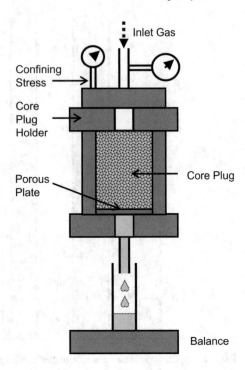

8.3   Capillary Pressure Determination

Capillary pressure is measured in core laboratory as a part of special core analysis
(SCAL) program. The laboratory experiments are performed on core plugs to establish capillary pressure saturation relationship that can be used in reservoir studies.
Mainly, three different experiments (methods) are available to estimate capillary
pressure in laboratory. These are the porous plate, the centrifuge and the mercury
injection methods. Out of these three methods, the porous plate is considered to be
the most reliable one.

8.3.1   Porous Plate Experiment (McCullough, 1944)

Figure 8.7 shows the porous plate (diaphragm) experiment. The porous plate used in
experiment is made of strongly water wet material (typically porcelain with fine pores
of uniform size. The experiment starts with fully saturating the core plug and the
porous plate with the wetting phase (water). They are kept in capillary contact inside
the core holder. Non-wetting phase (gas) pressure is increased slowly to reach the
threshold (displacement) pressure. Then, pressure is increased in predefined stepwise.
For each step, the pressure level is maintained until wetting phase flow stops.

Fig. 8.7 Capillary pressure determination centrifuge experiment

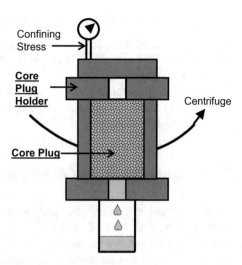

This method has the following advantages:

1. It is the most accurate method one and it is considered as a reference to other methods.
2. It can be used for heterogeneous and laminar samples.
3. No model is needed to interpret the experiment results.

In spite of being the most accurate, this experiment is not always preferred because it is very slow. Typically, each pressure level step requires a month to reach equilibrium. However, this disadvantage can be managed by the possibility of estimating the rock electric properties simultaneously with the capillary pressure.

8.3.2 Centrifuge Experiment (Hassler and Brunner, 1945)

Centrifuge method gives reasonably good results with shorter time and less cost than the porous plate method. This makes centrifuge method more attractive for some industry staff. Figure 8.7 shows the centrifuge experiment. Core plug is saturated with one fluid and in contact with the second one. The core plug holder is placed in a centrifuge. Centrifugal force is applied to generate a pressure gradient in each fluid. As the first fluid is produced from one plug end the second fluid enters from the opposite end to replace it. Centrifuge speed is then increased in steps and maintained constant at each step until production ceases. For each step, the cumulative production is recorded.

This method has the following advantages:

1. It is reasonably accurate.
2. It is much faster than the porous plate. The whole experiment can be finished in a matter of days or weeks.

The method has the following disadvantages:

1.   It is not accurate for heterogeneous and laminated samples.
2.   Mathematical calculation is needed to interpret the results. This may raise the possibility of errors.
3.   For highly permeable rocks, limited data may be reliably collected.
4.   It is not suitable for unconsolidated sands. The high speed of the centrifuge may cause sample damage.

8.3.3   Mercury Injection Experiment (Purcell, 1949)

For this method, extracted, dried and evacuated sample is first immersed in liquid mercury (Fig. 8.8). Mercury pressure is then increased in steps and the amount of mercury entering the sample is measured and converted to non-wetting phase saturation. As the gas phase (mercury vapor) pressure is very small, the recorded pressure (liquid mercury pressure) can be taken as the capillary pressure.

This method has the following advantages:

1.   It is the fastest method. The experiment can be completed in a matter of hours.
2.   Mercury is non-wetting phase to all known reservoir rock materials and under high pressure it can access the smallest rock pores.

Fig. 8.8 Capillary pressure determination mercury injection experiment

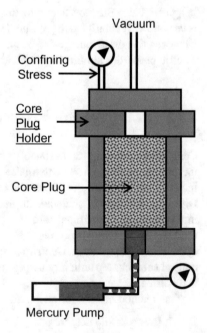

3. It is not necessary to have regular core plugs for this experiment. Rock samples of any shape or even drilling cuttings can be used. It is a common practice to use plug trim end for the experiment.
4. Mercury is non-wetting to all known reservoir rock materials and under high pressure it can access the smallest pores. Consequently, the experiment is excellent to study pore size distribution.

The method has the following disadvantages:

1. Mercury is not a reservoir fluid. The method is not expected to replicate the displacement process in the reservoir. Mercury injection capillary pressure is not recommended to be used for determining initial saturation distribution.
2. The experiment is destructive to the rock sample. A core plug used for this experiment cannot be used in other experiments.

8.4 Capillary Pressure, Lab to Rfeservoir Conversion

It can be recognized that the fluids used in capillary pressure experiments are not necessary the same fluids in the reservoir. Capillary pressure obtained from laboratory experiments need to be converted to represent the reservoir fluids before utilizing in reservoir studies. This conversion is performed by applying Eq. (8.2.2) for both laboratory and reservoir conditions.

For laboratory,

$$(P_c)_{lab} = (2\sigma \cos \theta / r)_{lab} \qquad (8.4.1)$$

For reservoir,

$$(P_c)_{res} = (2\sigma \cos \theta / r)_{res} \qquad (8.4.2)$$

Combining the above two formulae, the capillary pressure conversion factor can be simply written as,

$$\text{Conversion factor} = (\sigma \cos \theta)_{res} / (\sigma \cos \theta)_{lab} \qquad (8.4.3)$$

Tables 8.1, 8.2 and 8.3 explain the calculation of the capillary pressure conversion factor for different cases.

Tables 8.1, 8.2 and 8.3 show values of interfacial tension (σ), contact angle (θ) and capillary pressure conversion factor for different laboratory and reservoir fluid combinations. The values of σ and θ in these tables are the ones published by core laboratories (1982). Where available, more representative values should be used to estimate the conversion.

Table 8.1 Capillary pressure conversion factor for case-1

Laboratory			Reservoir			
Fluid system	σ (dyne/cm)	θ (°)	Fluid system	σ (dyne/cm)	θ (°)	Conversion factor
Brine gas	72	0	Brine oil	30	30	0.36084
Gas mercury	480	140	Brine oil	30	30	0.07066
Brine oil	48	30	Brine oil	30	30	0.62500
Core Laboratories, 1982			Core Laboratories, 1982			
Gas oil	24	0	Brine oil	30	30	1.08253
Dr. Paul Glover			Core Laboratories, 1982			

Table 8.2 Capillary pressure conversion factor for case-2

Laboratory			Reservoir			
Fluid system	σ (dyne/cm)	θ (°)	Fluid system	σ (dyne/cm)	θ (°)	Conversion factor
Brine gas	72	0	Oil gas	4	0	0.05556
Gas mercury	480	140	Oil gas	4	0	0.01088
Brine oil	48	30	Oil gas	4	0	0.09623
Core Laboratories, 1982			Core Laboratories, 1982			
Gas oil	24	0	Brine oil	4	0	0.16667
Dr. Paul Glover			Core Laboratories, 1982			

Table 8.3 Capillary pressure conversion factor for case-3

Laboratory			Reservoir			
Fluid system	σ (dyne/cm)	θ (°)	Fluid system	σ (dyne/cm)	θ (°)	Conversion factor
Brine gas	72	0	Brine gas	50	0	0.69444
Gas mercury	480	140	Brine gas	50	0	0.13598
Brine oil	48	30	Brine gas	50	0	1.20281
Core Laboratories, 1982			Core Laboratories, 1982			
Gas oil	24	0	Brine oil	50	0	2.0333
Dr. Paul Glover			Core Laboratories, 1982			

8.5   Pore Size Distribution

One of the important applications is to determine the pore size distribution using results from capillary pressure experiment results. The pore size distribution is a basic parameter for some reservoir characterization studies. It is a key factor for defining rock quality. It is also usable for calculating other rock parameters. As mentioned previously, the mercury is non-wetting phase to all known reservoir rocks materials. Under high pressure, the mercury can invade the smallest pores in the rock. Subsequently, the results from mercury injection experiment are excellent to determine the pore size distribution in a given rock sample (Burdine, 1953). The incremental saturation represents the relative volume of the specific pore size pores invaded by mercury. To calculate the specific pore size at any mercury saturation, Eq. (8.2.2) is applied considering the gas-mercury interfacial tension ($\sigma = 480$ dyne/cm) and the contact angle ($\theta = 140°$). The pore size can be calculated using one of the Eq. (8.5.1) or (8.5.2),

$$P_c = 2\sigma \cos \theta / r \tag{8.5.1}$$

$$r(\mu) = 7.35/P_c(bar) \tag{8.5.2}$$

$$r(\mu) = 106.6/P_c(psi) \tag{8.5.3}$$

where,

$r(\mu)$ pore size (radius), micron (μ m)
P_c(bar) capillary Pressure, bars
P_c(psi) capillary Pressure, psi.

The pore size distribution profile can be prepared as follows:

(1) Start with the capillary pressure saturation relationship as obtained from mercury injection experiment (Fig. 8.9).
(2) Convert the saturation axis to represent non-wetting phase (mercury) instead of wetting phase (gas) as in Fig. 8.10.
(3) Exchange the two axes so that the horizontal axis represents the capillary pressure and the vertical axis represents the non-wetting phase saturation (Figs. 8.11)

Fig. 8.9 Capillary pressure (mercury injection)

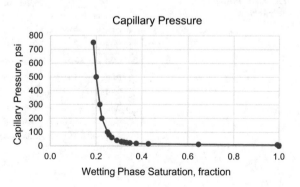

Fig. 8.10 Capillary pressure
(mercury injection)

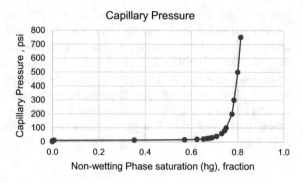

Fig. 8.11 Capillary Pressure
(Mercury Injection)

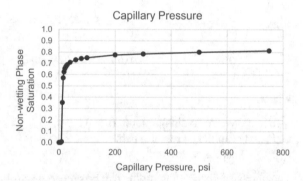

(4)   Convert the capillary pressure represented by the horizontal axis to pore size
       using Eq. (8.5.2) and replace the non-wetting phase saturation by the non-
       wtting phase incremental saturation (Fig. 8.12).

Figures 8.13 and 8.14 are alternative presentations of the pore size distribution profile.
The histogram presentation of Fig. 8.14 is the most preferred one.

Fig. 8.12 Pore size
distribution

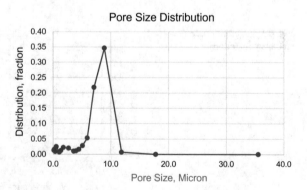

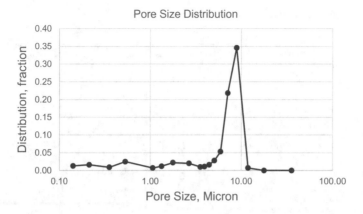

Fig. 8.13 Pore size distribution

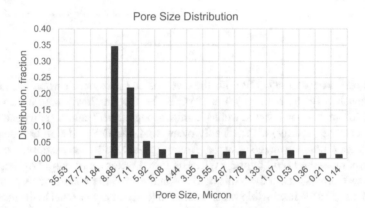

Fig. 8.14 Pore size distribution

8.6   Free Water Level, Oil Water Contact and Transition Zone

At initial reservoir condition, it is known that generally the water saturation increases with depth. This increase, related description of different levels and saturation zones are handled in this section. The connection of these parameters with the capillary pressure is also explained. In an oil reservoir the **free water level** (FWL) is defined as the level where both water phase pressure and oil phase pressure are equal ($P_{oil} = P_{water}$). In other words, it is the level where water oil capillary pressure is equal to zero. The FWL is determined by the intersection of the water and oil pressure gradients. Pressure data from MDT or similar measurements at initial reservoir condition are used for this purpose (Fig. 8.15).

Fig. 8.15 Free water level &
oil water contact

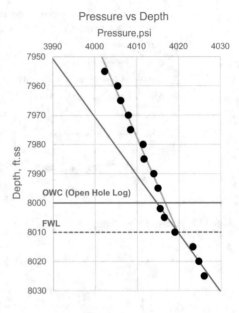

The **oil water contact** (OWC) is the level inside the reservoir where water satu-ration reaches 100%. Normally, the OWC is few feet shallower than the FWL. In the case of having one or more wells passing through the oil column and penetrating the water zone (aquifer) in reasonably clean sand, the OWC is simply recognized through open-hole log analysis. In the absence of such wells, it is a common practice to approximate the OWC by the FWL. The initial water saturation in the reservoir decreases from its maximum value of 100% at the oil water contact to its minimum value defined as the irreducible water saturation (S_{wirr}) at higher level in the reservoir. The critical water saturation (S_{wcr}) is the minimum water saturation value at which the water phase is mobile. This value is higher than the irreducible water saturation and expected to be at some deeper depth. For most practical applications, the two saturations are equal and the water phase starts to be mobile at S_{wirr}.

The **transition zone** is the zone extending from the level of the critical water saturation (top of transition zone, TTZ) to the level of the OWC. In this zone, the two phases (oil and water) are mobile. Figure 8.16 is a schematic illustration to show initial water saturation distribution, FWL, OWC and TTZ. In this figure the bottom of transition zone is considered at the OWC level. However, some scholars prefer to be strict about defining the transition zone as the zone where both oil and water are mobile. Accordingly, they consider the bottom of the transition zone (BTZ) at the level where oil reaches its residual saturation (S_{orw}) (Fig. 8.17).

The height above the FWL can be converted to capillary pressure (Eq. 8.3.2) and vice versa. It can be recognized from the Fig. 8.16 or 8.17 that the relation between the capillary pressure and saturation can be utilized to determine the initial water saturation in the reservoir. Figure 8.18 for more clarification about the relation between water saturation and capillary pressure/height above free water level in the

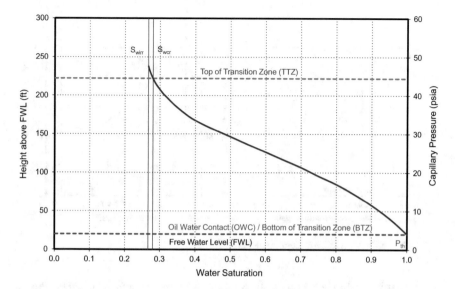

Fig. 8.16 FWL, OWC and transition zone

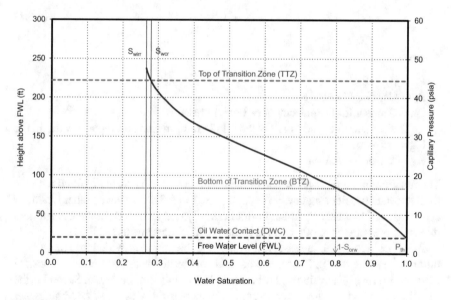

Fig. 8.17 FWL, OWC and transition Zone

reservoir. It is to be insisted that this relation is only applicable in the transition zone and at the initial reservoir condition. Initial water saturation shallower than the top of transition zone is constant at its minimum value (S_{wirr}). Obviously, water saturation deeper than the OWC is constant at its maximum value (100%). In the transition

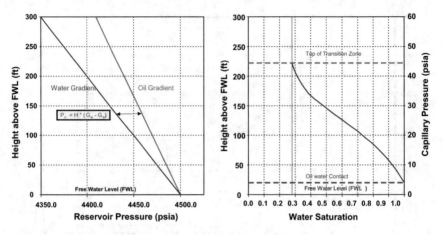

Fig. 8.18   Capillary pressure & fluid gradients

zone, the relationship between capillary pressure and water saturation can be written as;

$$P'_c = f(S_w) \qquad (8.6.1)$$

where,

$P'_c = P_c - P_{th}$
P_c = Capillary Pressure
P_{th} = Threshold (Displacement or Entry) Pressure
    The pressure needed to allow the oil phase to invade and replace water in the biggest size pores
S_w = Water Saturation.

The function in the right-hand side of the Formula (8.6.1) is normally fitted with the available data considering power function $\left[f(S_w) = a\, S_w^b \right]$. However, other functions can be tried. Obviously P'_c in the Formula (8.6.1) will turn to P_c in the case of zero threshold pressure or in other words if the OWC is the same as the FWL.

In this consequence, the reservoir system can be initially divided into three saturation zones (Fig. 8.19). First is the oil zone where the water saturation is constant at its minimum (S_{wirr}) value and oil is initially the only mobile phase. Second is the transition zone where the water saturation is increasing with depth and both water and oil are mobile. Third is the water zone (aquifer) where water saturation is at its maximum (100%) value and only water is mobile. Referring to the same figure, well W$-$1 will initially produce water free oil, well W$-$2 should be completed in the oil zone to avoid early water production and well W$-$3 will most probably produce water with oil at start. It is well known that water encroachment is a gas reservoir

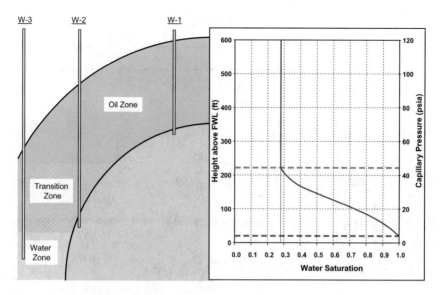

Fig. 8.19 Initial saturation zones

killer. Subsequently, careful selection of well completion to avoid watering the wells is more crucial.

In the above discussion, the reservoir has been considered as homogenous. For such reservoir, the initial water saturation-depth relation can be handled with a single capillary pressure curve and the water saturation is increasing with depth. This is not the normal case. Figure 8.20 illustrates a water saturation profile along a newly discovered oil reservoir. From this figure, the role of initial water saturation increasing with depth is not fully valid. By further studying the reservoir rock, it has been recognized that it consists of a number of layers with three different rock types represented by three capillary pressure curves (Figs. 8.21, 8.22 and 8.23).

8.7   Capillary Pressure Averaging

8.7.1   Leverette J-Function

In our reservoir engineering work, there is a need of averaging or smoothing the available capillary pressure curves coming from different experiments. The famous **Leverette** (1941) **J-Function** or simply the J-Function is the most popular way to do this averaging. The J-Function was originally proposed aiming to develop a universal curve for all capillary pressure data. That was not possible due to the significantly different capillary pressure for different rock types. Knowing this the use of this function for capillary pressure averaging is preferred to be limited to data of the

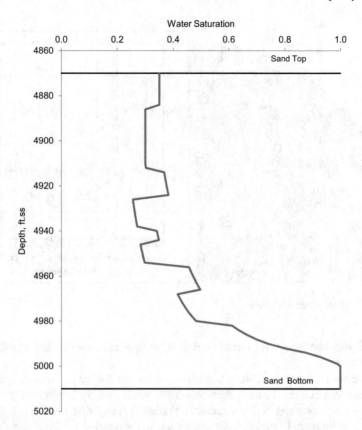

Fig. 8.20 Reservoir layers with different rock types

same rock type. However, some scholars use the J-Function to average capillary pressure of whole under study reservoir regardless of rock types. This could result in over approximation.

The J-Function (J) is defined by the formula,

$$J = P_c / \sigma \cos \theta (K/\Phi)^{0.5} \qquad (8.7.1.1)$$

where,

J = J-Function
P_c = capillary pressure, dyne/cm^2
K = permeability, cm^2
Φ = porosity, fraction
σ = interfacial tension, dyne/cm
θ = contact angle.

Fig. 8.21 Reservoir layers
with different rock types

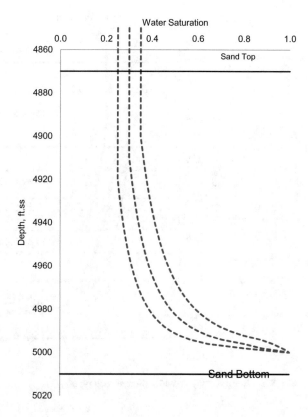

Dimension analysis of the above formula shows that the J-Function is a dimensionless quantity and it can be simply described as the dimensionless capillary pressure. In the Formula (8.7.1.1), the CGS unit system is used. For practicality, the formula can be re-written in field units as,

$$J = 0.217 P_c / \sigma \cos \theta (K/\Phi)^{0.5} \tag{8.7.1.2}$$

where,

$J = $ J-Function
$P_c = $ capillary pressure, psi
$K = $ permeability, md
$\Phi = $ Porosity, fraction
$\sigma = $ Interfacial Tension, dyne/cm
$\theta = $ Contact Angle.

Using anyone of the above two formulae will calculate the same dimensionless J-Function. This also applies to any similar formula considering different unit system as long as this unit system is carefully respected.

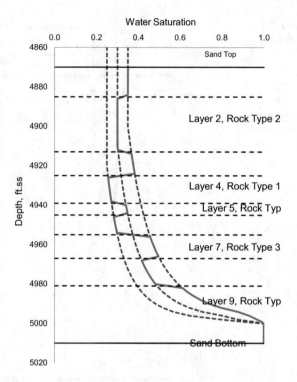

Fig. 8.22 Reservoir layers with different rock types

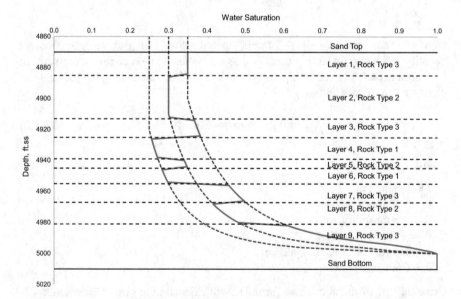

Fig. 8.23 Water saturation profile, rock types

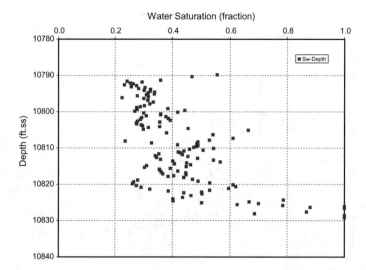

Fig. 8.24 Pseudo capillary pressure water saturation versus depth (sw profile)

8.7.2 Water Saturation-Depth Profile, Pseudo Capillary Pressure

Another method to create average capillary pressure is to prepare the so-called water saturation-depth profile or water saturation profile (S_w Profile). This is performed by plotting open-hole log evaluated water saturation vs. subsea depth. For ideal case, the curve fitting the plotted points will represent the increase of water saturation with depth. The curve can be simply converted to average capillary pressure (Eq. 8.2.3) relation which represents the reservoir (Drake, 2001). Figure 8.24 shows water saturation-depth plot for an oil reservoir. The scattering of the plotted points can be overcome by splitting data into different rock groups. For this example, the data have been split according to different porosity bins (Fig. 8.25). As can be noticed, it is possible to end up with S_w Profile for each rock group. It is obvious that this S_w Profile can be converted to capillary pressure-water saturation relationship. This method of creating the capillary pressure-water saturation relationship has the advantage of considering real water saturation data. Of course, this is limited by the reliability of the water saturation evaluation. Additionally, it is worth mentioning that this method can only be reliable in the case of good coverage of well logs all over the reservoir.

8.8 Saturation Height Function (SHT)

The saturation height function (SHF) is a function generated using the available capillary pressure curves with other reservoir parameters aiming at creating field wide initial water saturation distribution. Various saturation height functions are

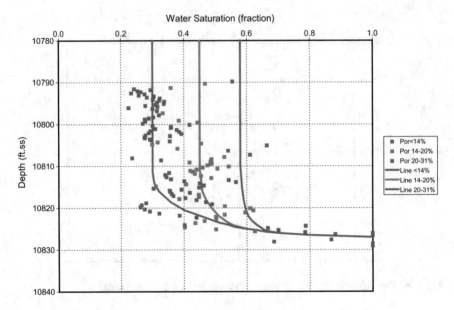

Fig. 8.25 Pseudo capillary pressure water saturation versus depth (sw profile)

available. Table 8.4 list the most applicable saturation height functions (Harrison and Jing, 2001).

The classical Leverette (1941) SHF technique is the most popular one. The technique relies on considering a power relationship between capillary pressure and water saturation. Similar power relationship can also be considered between J-Function and the water saturation. This technique is described in the following:

(1)   Laboratory capillary pressure data for the similar rock type are converted to J-Function (Eq. 8.7.1.2). This step results in J-Function water saturation curve for each capillary pressure water saturation curve from laboratory.

(2)   The J-Function curves are averaged in one J-Function curve representing one rock type. The representative J-Function curve is utilized to determine the two constants a and b in the formula:

$$J = a(S_w)^{-b} \qquad (8.8.1)$$

(3)   Capillary pressure (P_c) at any depth can be calculated using equation:

$$P_c = h(G_W - G_o) \qquad (8.8.2)$$

Table 8.4 Capillary pressure conversion factor for case-4

SHF	Formula	Notes
Leverette (1941)	$J = a(S_w)^{-b} \rightarrow S_w = (a/J)^{1/b}$ $S_w =$ $\left[(a/0.217)(\sigma \cos \theta/P_c)(\Phi/K)^{0.5}\right]^{1/b}$	$S_w =$ water saturation, fraction $J = $ J-Function $P_c = $ capillary pressure, psi $\sigma = $ interfacial tension, dyne/cm $\theta = $ contact angle $\Phi = $ porosity, fraction $K = $ permeability, md a and b constants (found by regression)
Johnson (1987)	$\text{Log}(S_w) = BP_c^{-C} - A \, \text{Log}(K)$	$S_w = $ water saturation, % $P_c = $ capillary pressure, psi A, B and C constants (derived from SCAL P_c data)
Cuddy et al. (1993)	$\text{Log}(\Phi S_w) = A \, \text{Log}(h) + B$	$S_w = $ water saturation, fraction $\Phi = $ porosity, fraction h = height above FWL, ft A and B constants (found by regression)
Skelt Harrison (1995)	$S_w = 1 - A \, \exp\left[-\left(\frac{B}{h+D}\right)^c\right]$	$S_w = $ water saturation, fraction $\Phi = $ porosity, fraction h = height above FWL, ft A, B, C and D constants (found by regression)

(4)   Water saturation is then calculated using:

$$S_w = \left[(a/0.217)(\sigma \cos \theta/P_c)(\Phi/K)^{0.5}\right]^{1/b} \quad (8.8.3)$$

where,

$P_c = $ capillary pressure, psi
h = height above FWL, ft
$G_w = $ water pressure gradient, psi/ft
$G_o = $ oil pressure gradient, psi/ft
$\sigma = $ water oil interfacial tension, dyne/cm
$\theta = $ contact angle
$\Phi = $ porosity, fraction
$K = $ permeability, md.

The above procedure can be repeated for each rock type in the reservoir. However, some industry people prefer to make one function for the whole reservoir regardless of different rock types. This simplification should be avoided where we have quite heterogeneous reservoir.

It is to be mentioned that water saturation is a function of depth only in transition zone at initial reservoir condition. Above the transition zone, only irreducible water initially exists ($S_w = S_{wirr}$). Below the transition zone, water saturation is at its maximum value ($S_w = 1.0$). Accordingly, using SHF to estimate initial water saturation distribution is limited in transition zone at initial reservoir condition.

8.9   Reservoir Simulation Model Initialization

It is understood that the reservoir simulator basically solves for pressure and saturation for each reservoir model block at different time steps. To do this, the simulator primarily needs to have the saturation and pressure distribution at initial condition (zero time). Two sets of data are used to fulfill this purpose. First set includes initial reservoir pressure at datum and fluid contacts' levels. This set of is provided through one or more equilibration card. As an example to explain the procedure, let us consider a reservoir whose initial pressure is 2900 psia @ 5800 ft.ss, the oil water contact (OWC) is at 5900 ft.ss. and the gas oil contact (GOC) is at 5000 ft.ss. For simplicity, the OWC level is assumed to be the same as the FWL. The equilibration card is provided as,

EQUIL

5800 2900 5900 0.0 5000 0.0 /

The second set of data includes the capillary pressure-water saturation relationship information given as tabular form of capillary pressure values versus water saturation values. This form of data will be more discussed in Chap. 9 (Saturation Functions).

The procedure followed by the simulator to set initial pressure and saturation values all over the reservoir is illustrated in Figs. 8.26 through 8.30 and described below:

(1)   The simulator sets the initial oil phase pressure at datum as provided in the initial equilibrium information (Fig. 8.26). The initial oil phase pressure for any reservoir model block whose center is at the same depth as the given datum will be set as the given initial pressure.

(2)   The simulator calculates the initial oil phase reservoir pressure for other blocks using the oil phase pressure gradient which can be estimated from the given PVT information (Fig. 8.27). In this step the model also sets the water phase pressure at the OWC. Note that the oil phase pressure and water phase pressure are equal at FWL and it is assumed that the OWC and FWL levels are same.

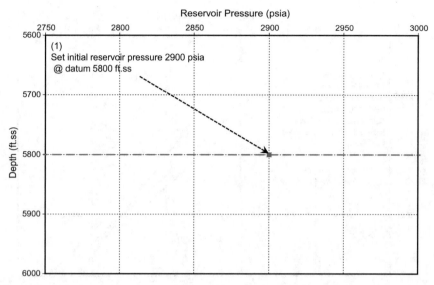

Fig. 8.26 Simulation model initialization

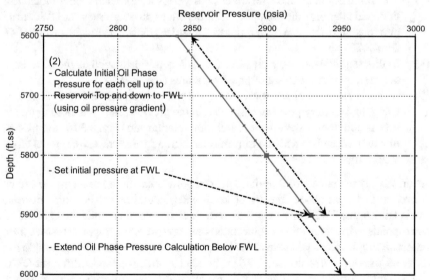

Fig. 8.27 Simulation model initialization

EQUIL
5800 2900 5900 0.0 5000 0.0 /

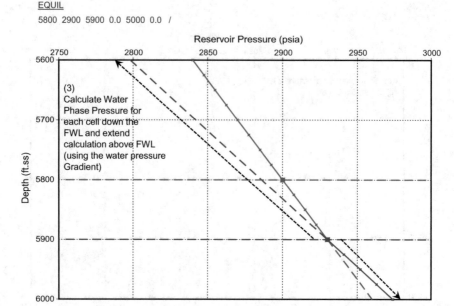

Fig. 8.28 Simulation model initialization

(3)   The simulator calculates the water phase pressure for each model block. For these calculations, the simulator uses the water phase pressure at OWC from previous step and the water phase pressure gradient estimated from PVT (Fig. 8.28).

(4)   In this step, the capillary pressure for each cell is estimated as the difference between oil phase and water phase pressures (Fig. 8.29).

(5)   In this last step, the simulator set the initial water saturation value for each block using the capillary pressure estimated in the previous step and looking up the given capillary pressure-water saturation relationship (Fig. 8.30). Completing this step, we end up with pressure and saturation for each reservoir model block at zero time.

As previously explained, the capillary pressure-water saturation relationship can be converted to height above free water level-water saturation relationship. In these two relations, the FWL corresponds to zero capillary pressure while the OWC level corresponds to the threshold (displacement or entry) pressure. Suppose that we have different samples from what we know to be the same reservoir and each sample has different threshold pressure (Fig. 8.31). In other words, we expect different OWC levels. This information may guide to considering different reservoir blocks. This is not necessary right. Different OWC levels can exist in the same reservoir. Actually, the OWC is not necessary a horizontal plane as commonly understood. It can take the shape of surface (Fig. 8.32). This is especially noticed in heterogeneous reservoirs with relatively big aerial extension. In such case, it seems that the simulator should be

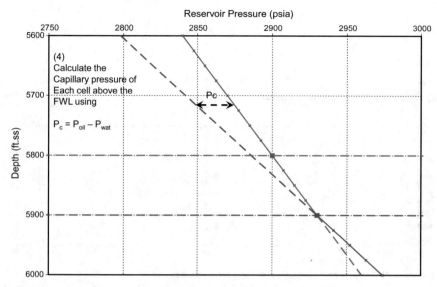

Fig. 8.29   Simulation model initialization

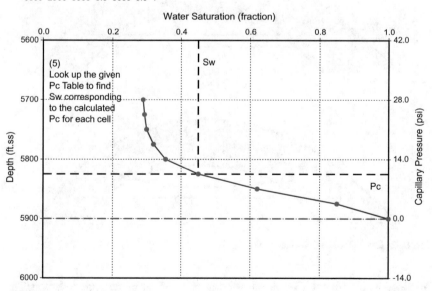

Fig. 8.30   Simulation model initialization

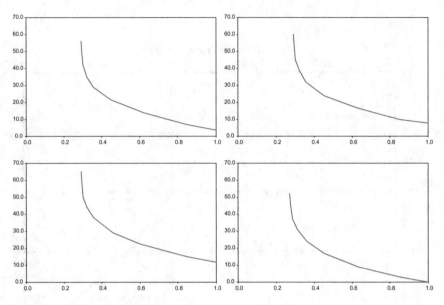

Fig. 8.31   Different threshold (displacement, entry) pressure

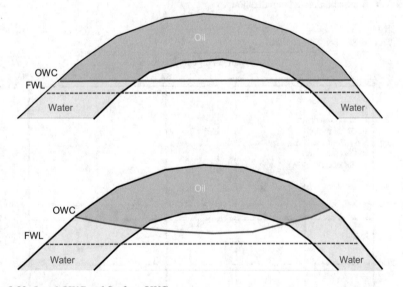

Fig. 8.32   Level OWC and Surface OWC

provided with different OWC levels for different regions in the same reservoir. This could result in initial model instability. To avoid this, it is suggested to provide the simulator with the FWL as the common OWC. Additionally, to modify the available capillary pressure-water saturation data so that the water saturation value at the

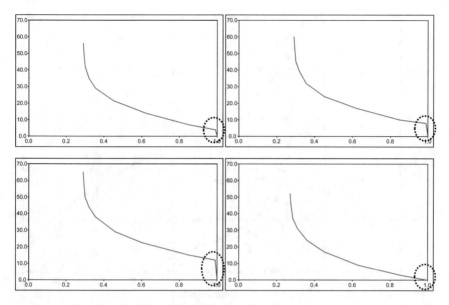

Fig. 8.33 Different threshold (displacement, entry) pressure

threshold pressure is reduced from its original value of 1.0 to slightly lower value. Figure 8.33 where the water saturation at threshold pressure was reduced from 1.0 (100%) to 0.99 (99%). This modification would cause change in calculated fluid volumes but it obvious that this change is expected to be insignificant.

8.10 Drainage and Imbibition Capillary Pressure Curves

In our discussion so far, we consider the basic drainage capillary pressure curve which represents the process of oil migration to the reservoir and used to distribute the water saturation in the initial reservoir static system. In addition to the basic drainage curve, Fig. 8.34, shows another capillary pressure curve. This curve is the imbibition capillary pressure curve which represents the process of oil depletion. The imbibition data can be useful in assessing the relative contributions of capillary and viscous forces in dynamic system. These two curves are practically good enough for modeling reservoir performance. For many SCAL programs, only the first (drainage) curve is requested. It was found that the absence of the second curve (imbibition) is not crucial and does not have significant impact on simulation results for most cases. Other possible combinations of capillary pressure curves are shown in Figs. 8.35, 8.36 and 8.37.

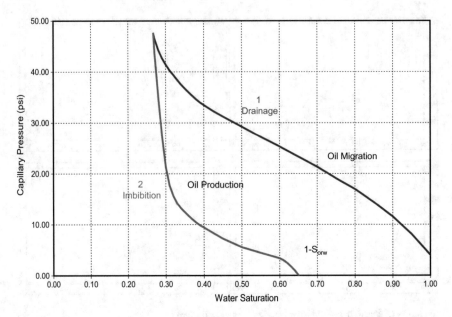

Fig. 8.34 Drainage and imbition capillary pressure

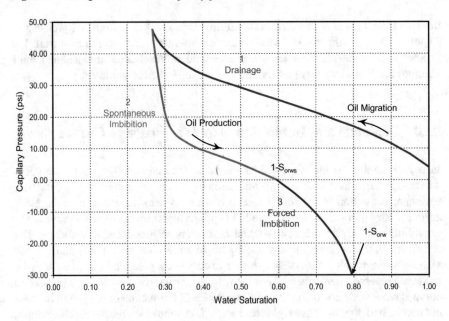

Fig. 8.35 Drainage and imbition capillary pressure

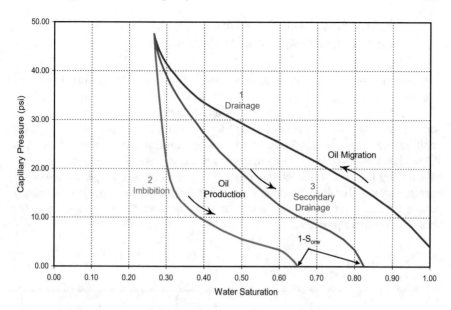

Fig. 8.36 Drainage and imbibition capillary pressure

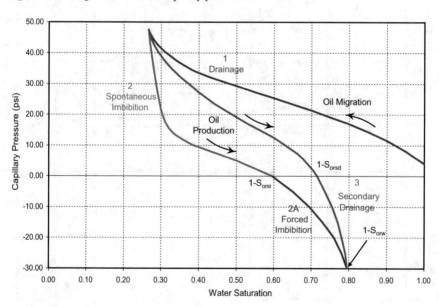

Fig. 8.37 Drainage and imbibition capillary pressure

8.11   Gas Oil Capillary Pressure Curve-Oil Reservoir and Gas Cap Saturation Distribution

In the whole previous discussion, water oil system is assumed. The capillary pressure, in this discussion, simulates the saturation distribution in the water oil transition zone. The question is what about the gas oil system. Looking back at Tables 8.1 and 8.2, it is easy to conclude that the gas oil capillary pressure is about 15% of the water oil capillary pressure. Figure 8.38 shows a typical gas oil capillary pressure curve. This figure shows how the gas oil capillary pressure is compared in value to the water oil capillary pressure. It is a common practice to assume that the gas oil transition zone is very small or does not exist. This leads to approximating the gas oil capillary pressure to zero. It is important to insist that this approximation is not applicable for water gas capillary pressure.

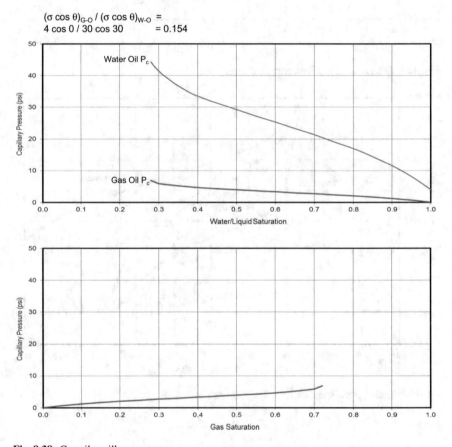

Fig. 8.38 Gas oil capillary pressure

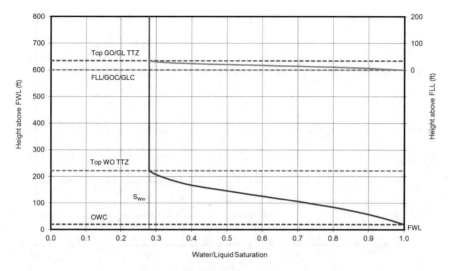

Fig. 8.39 Water–oil and gas-oil transition zones

The plot of Fig. 8.39 is a schematic presentation of the initial water saturation distribution in the whole reservoir system (including the gas cap). From this plot, the definitions of the FWL, OWC and Water Oil TTZ (WO TTZ) stand as previously explained. The free liquid level (FLL) is the level where the gas oil capillary pressure is equal to zero or in other words the gas phase pressure and oil phase pressure are the equal. The gas oil contact (GOC) is the shallowest level in the system where gas does not initially exist ($S_g = 0$) or in other words where the liquid (water + oil) saturation is 100%. The gas liquid contact (GLC) has the same definition. In the given plot the FLL and the GOC are on the same level which means that the gas-oil system threshold pressure is equal to zero. The gas oil top of transition zone (GO TTZ) is the shallowest level where there is no liquid hydrocarbon ($S_o = 0$). Considering the practical assumption that the gas oil transition zone does not exist, the schematic presentation will change to the one in Fig. 8.40.

8.12 Capillary Migration

Figure 8.41 shows a chain of pore through with a small amount of oil engaged in moving over a pore network which was originally filled with water. This oil drop challenges to rise, and its face H is Infront of a bottleneck with the radius "r". So, to pass this bottleneck, it is essential that the oil–water contact surface take sphere shape with radius "r".

Therefore, in H on each side of the pore thought there is difference in pressure, P_{Hw} and P_{Ho}:

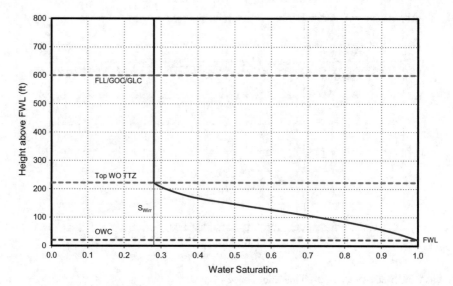

Fig. 8.40 No gas oil transition zone

Figure 8.41 Displays pore through with a small amount of oil engaged in moving over a pore network

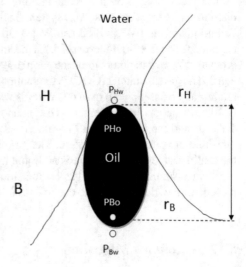

$$P_{Ho} - P_{Hw} = 2T/r_H \qquad (8.12.1)$$

The base of the oil drop is at point B and has a circular shape with the radius "r_B". Similarly, in the water phase there are P_{Bw} and P_{Hw}:

$$P_{Bw} = P_{Hw} + h\rho_w g \qquad (8.12.2)$$

where

h = height of the oil drop,
ρ_w = water density,
g = gravitational acceleration.

At point B in the oil phase there are P_{Bo} and P_{Ho}:

$$P_{Bo} = P_{Ho} + h\rho_o g \tag{8.12.3}$$

where ρ_o is oil density,

At point B on each side of the pore thought there is:

$$P_{Bo} - P_{Bw} = 2T/r_B \tag{8.12.4}$$

Hence the following is deduced:

$$2T/r_b = \left(P_{H_o} - P_{Hw}\right) - h(\rho_w - \rho_o)g$$

Then

$$P_{Ho} - P_{Hw} = h(\rho_w - \rho_o)g + 2T/r_B \tag{8.12.5}$$

If $P_{Ho} - P_{Hw} = h(\rho_w - \rho_o)g > 2T/r_B$ the oil drop will go over the pore through.
If $P_{Ho} - P_{Hw} = h(\rho_w - \rho_o)g < 2T/r_B$ the oil drop will not go over the pore through.

If assumed that $r_H = 0.5\ \mu$ and $T = 30$ dyn/cm, then:

$$2T/r_B = (2 \times 30)/\left(0.5 \times 10^{-4}\right) = 1.2\ \text{bar}$$

which is a quite high value.

8.13 Effects of a Pressure Gradient

The oil drops are subject to the buoyancy and the drive of the medium. It is simple to illustration that the oil accumulation does not necessarily go on at the high point of the pore through and that the point of the oil accumulation changes from horizontal to the slope dz/dx such as follows:

$$dz/dx = d_{hw}/dx/\left(((\rho_w - \rho_o)/\rho_w) - (d_{hw}/dz)\right) \tag{8.13.1}$$

where h_w is the hydrostatic column of the water. Hydrocarbons can accumulate in many pores with due to different buoyancy. The inclination of the base of hydrocarbons accumulation is investigated very deeply in basin hydrodynamics. These notional facts are rarely confirmed.

Exercises:

1. The capillary pressure for a capillary tube with circular cross section is given by the formula:

$$P_c = 2\sigma \cos \theta / r$$

What would be the formula for a capillary tube with square cross section?

2. The given table includes mercury injection capillary pressure data for two samples as reported. If these samples were taken from an oil reservoir.

Sample ID	Depth (ft)	Press (psi)		3	6	9	12	15	18	21	24	27
		K(md)	Φ (%)	Wetting phase saturation (%)								
S-01	9250	622	24	100	100	99.3	64.7	42.9	37.6	34.8	33.2	32.1
S-02	9265	2070	25	97.9	51.9	37.1	31.7	28.5	27.0	25.4	24.0	23.2

Sample ID	Depth (ft)	Press (psi)		30	40	60	80	100	200	300	500	750
		K(md)	Φ (%)	Wetting phase saturation (%)								
S-01	9250	622	24	31.1	29.1	26.9	25.7	25.0	22.5	21.6	20.0	18.7
S-02	9265	2070	25	22.4	20.3	18.1	17.0	16.2	13.8	12.5	11.1	9.6

(a) If this samples were taken from an oil reservoir, convert the given data to reservoir condition. Plot the obtained results and give comments.
(b) Perform similar conversion if the samples are from dry gas reservoir.
(c) Prepare pore size distribution profiles for the two samples. Comment on the results.
(d) Apply Leverette SHF technique to calculate the initial water saturation at two points in the reservoir 50 and 100 ft above the FWL. Water pressure gradient and oil pressure gradients are 0.450 psi/ft and 0.280 psi/ft respectively.

References

Burdine, N. T. (1953). Relative permeability calculation from pore size distribution data. *Transactions on AIME 198*, 71–78.

Cuddy, S., Allinson, G., & Steele, R. (1993). A simple convincing model for calculating water saturations in southern North Sea gas fields. SPWLA 34th Annual Logging Symposium, June 13–16.

Drake, L. P. (2001). *Fundamentals of reservoir engineering*. Elsevier Science B.V. Amsterdam, The Netherlands.

Harrison, B., & Jing, X. D. (2001). Saturation height methods and their impact on volumetric hydrocarbon in place estimates. In: *Society of petroleum engineers, SPE 71326*. https://doi.org/10.2118/71326-MS.

Hassler, G. I., & Brunner, E. (1945). Measurement of capillary pressures in small core samples. *Transactions on AIME, 160*, 114–123.

Johnson, D. L., Koplik, J., & Dashen R. (1987). Theory of dynamic permeability and tortuosity in fluid-saturated porous media. *Journal of Fluid Mechanics, 176*, 379–402.

Leverett, M. C. (1941). Capillary behaviour in porous solids. *Transactions of the AIME, 142*, 159–172.

McCullough, J. J. (1944). Determination of the interstitial-water content of oil and gas sand by laboratory tests of core samples. API-44–180.

Purcell, W. R. (1949). Capillary pressures-their measurement using mercury and the calculation of permeability thereform. *Transactions on AIME, 186*(3), 39–48.

Skelt, C., Harrison B. (1995). *An integrated approach to saturation height analysis*. SPWLA 36th Annual Logging Symposium, June 26-29.

Tarek, G. (2019). *Fundamentals of reservoir rock properties*. Springer International Publishing. ISBN 978-3-030-28139-7.

Chapter 9
Relative Permeability

9.1 Relative Permeability

A porous medium saturated with different fluids has different effective permeability to each one fluid. **Relative permeability** to one phase is the dimensionless measure of the effective permeability to that phase in multiphase flow in a porous medium (Mandal, 2007). For any one phase, the relative permeability is evaluated by dividing the effective permeability to that phase by the absolute permeability of the porous medium,

$$K_{r-ph} = K_{e-ph}/K_{abs} \tag{9.1.1}$$

It was found that the relative permeability to one phase changes with the saturation of that phase. It is equal to one at 100% saturation of the phase and gradually decreases to reach zero at the critical or irreducible saturation of that phase. In any hydrocarbon reservoir system, no one phase can reach the saturation of 100%. Consequently, in a multiphase system, the relative permeability to any phase cannot reach the value of one. Figure 9.1 is a typical relative permeability plot describing the relative permeability-water saturation relationship for a water–oil system. Figure 9.2 is a similar plot for a gas-oil system. It is clear that the water–oil system is a two phase system (water and oil). The gas-oil system is also considered as a two phase system. However, this system actually contains water at constant saturation (irreducible water saturation). Some reservoir engineering scholars prefer to define it as gas-oil-irreducible water system.

In most core analysis laboratories, the reported relative permeability is calculated by referencing the effective permeability to any phase to the effective permeability to oil at irreducible water saturation. Table 9.1 is a typical SCAL report; refer to the foot note of the table. In other words, the absolute permeability (K_{abs}) as reference permeability in the above Eq. (9.1.1) is replaced by the maximum effective permeability to oil (maximum K_o or K_o @ S_{wirr}). This practice results in calculating a

© The Author(s), under exclusive license to Springer Nature Switzerland AG 2022
A. M. Badawy and T. A. A. O. Ganat, *Rock Properties and Reservoir Engineering: A Practical View*, Petroleum Engineering,
https://doi.org/10.1007/978-3-030-87462-9_9

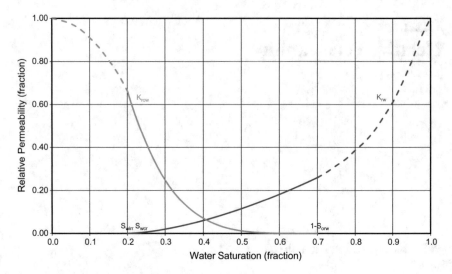

Fig. 9.1 Water oil relative permeability

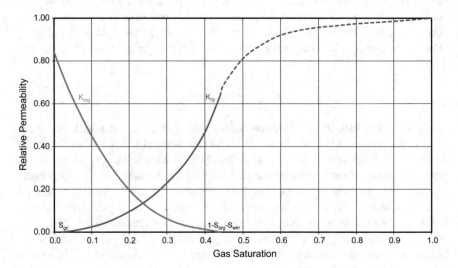

Fig. 9.2 Gas oil relative permeability

maximum relative permeability of 1.0 to the oil phase. Note the first entry of the column of the relative permeability to oil in the same table. It is necessary to adjust all reported relative permeability values to be referenced to the absolute permeability before further data proceeding.

The adjustment factor and the adjusted relative permeability can simply be calculated by,

Table 9.1 Water–oil relative permeability: SCAL report

Well: 6G-1
Sample identification:3c
Sample depth, Meters: 1247.33
Permeability to Air: 406 md
Porosity: 33.3%
Initial water saturation 20.6%
Effective permeability to oil at S_{wi}: 397 md

Water saturation, percent pore space	Water–oil relative permeability ratio	Relative permeability to water, fraction	Relative permeability to oil, fraction
20.6	0.0	0.0	1.0
34.1	0.177	0.072	0.409
36.0	0.227	0.084	0.371
39.0	0.357	0.10	0.279
59.0	13.2	0.275	0.021
60.6	19.0	0.297	0.016
61.7	24.8	0.313	0.013
63.2	36.2	0.327	0.009
64.2	49.9	0.355	0.0071
65.2	67.1	0.369	0.0055
66.4	93.7	0.390	0.0042
68.9	301	0.422	0.0014
70.1	–	0.449	–

$$\text{Adjustment Factor (F)} = (K_o @ S_{wirr})/K_{abs} \tag{9.1.2}$$

$$K_r \text{ (adjusted)} = F^* K_r \text{ (reported)} \tag{9.1.3}$$

For dry gas reservoirs, the water gas system normally replaces the water oil system experiment in the SCAL program. In this case the reference permeability used in reporting relative permeability data is accordingly changed to be the maximum effective permeability to gas (maximum K_g or $K_g @ S_{wirr}$). The adjustment factor should be changed accordingly.

It can be noticed that the concept of the relative permeability is quite simple. However, proper evaluation is not an easy task. The relative permeability is evaluated in laboratory as a part of the SCAL program work. Steady state displacement, unsteady state displacement and centrifuge experiments are used to evaluate relative permeability at different saturation values in laboratory. Steady state experiments are generally preferred as a relative permeability evaluation methodology. The high cost of this experiment makes it necessary to optimize the SCAL requirements. A general practice is to perform unsteady state relative permeability experiments for most of the selected plugs and keep few plugs for steady state experiments as a reference.

The centrifuge experiment is used for limited purpose (relative permeability end points). Other methods of evaluating the relative permeability include well testing and correlation (Ramli et al., 2011). As a result of well test analysis, the permeability is one of the coming results. At the early reservoir life, this test permeability is really the average effective permeability to the flowing phase (oil or gas). Knowing the absolute permeability of the rock, it is possible to evaluate the relative permeability. Considering this, a single relative permeability-saturation point can be evaluated with some limitation. The concept looks fine but it is really unpractical.

It is a common practice to use correlation to evaluate the relative permeability for a reservoir with no or limited SCAL data. A relative permeability-saturation mathematical relationship can be established from available SCAL data of analogue reservoir/s. Subsequently, such relationship can be used to evaluate the relative permeability for other reservoirs with no or poor SCAL data. The most popular correlation method is the Corey Type correlation. For this correlation type, a power function is suggested to relate the relative permeability to a phase with its saturation. The general form of this correlation formula is,

$$K_r^* = \left(S^*\right)^N \tag{9.1.4}$$

where,

K_r^* is the normalized relative permeability to phase.
S^* is the normalize saturation of the phase.
N   is known as Corey exponent and it represents the shape of the relative permeability-saturation curve.

Before going to more details about relative permeability correlation, let us clarify about **normalization**, how it is to be performed and why it is needed. Normalizing a parameter S means to express it as a fraction of a desired or practical range. In Fig. 9.3, the parameter S has the possible values ranging from 0 to S_{max}. However, its practical range is between S_1 and S_2. It is desired to express the parameter S as a fraction of its practical range (S_1 to S_2). The normalized parameter can then be calculated as,

$$S^* = (S - S_1)/(S_2 - S_1) \tag{9.1.5}$$

Doing this, the normalized value of the parameter (S^*) ranges from 0.0 to 1.0 as its real value ranges from S_1 to S_2. Vice versa, **de-normalization** is to estimate a parameter from its normalized information. The de-normalized S is calculated as,

Fig. 9.3 Normalization

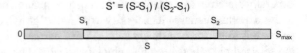

$$S = S^*(S_2 - S_1) + S_1 \tag{9.1.6}$$

Normalization helps to effectively combine, compare and average different groups of data.

Let us now go back to the relative permeability-water saturation relationship as illustrated in Fig. 9.4 (water oil system). The relative permeability to water ranges theoretically from 0.0 to 1.0 which corresponds to range of water saturation from 0.0 to 1.0 as well. However, the practical range of relative permeability to water is from 0.0 to a maximum value (K_{rw} @ $1-S_{orw}$) which corresponds to water saturation range from S_{wcr} to $1-S_{orw}$. Considering this, the normalized water saturation and the normalized relative permeability to water can be calculated as;

$$S_w^* = (S_w - S_{wcr})/(1 - S_{orw} - S_{wcr}) \tag{9.1.7}$$

$$K_{rw}^* = K_{rw}/(K_{rw} @ 1 - S_{orw}) \tag{9.1.8}$$

Similarly, the relative permeability to oil ranges theoretically from 0.0 to 1.0 which corresponds to range of oil saturation from 0.0 to 1.0 as well. The practical range of relative permeability to oil is from 0.0 to a maximum value (K_{ro} @ S_{wirr}) which corresponds to oil saturation range from S_{orw} to $1-S_{wirr}$. Considering this, the normalized oil saturation and the normalized relative permeability to oil can be calculated as;

$$S_o^* = (S_o - S_{orw})/(1 - S_{wirr} - S_{orw}) \tag{9.1.9}$$

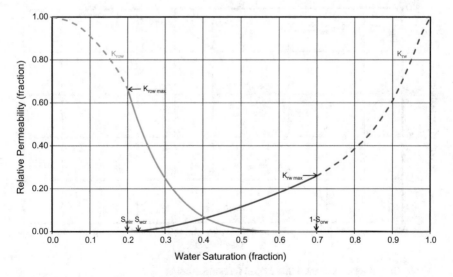

Fig. 9.4 Water oil relative permeability ranges for normalization

$$K_{ro}^* = K_{ro}/(K_{ro}@S_{wirr}) \tag{9.1.10}$$

Knowing that ($S_o = 1-S_w$), the Eq. 9.1.9 can be re-written as;

$$S_o^* = (1 - S_w - S_{orw})/(1 - S_{wirr} - S_{orw}) \tag{9.1.11}$$

Note that in the case of ($S_{wcr} = S_{wirr}$),

$$S_o^* = 1 - S_w^* \tag{9.1.12}$$

Due to the existence of a third fluid (irreducible water), the analysis of gas oil system relative permeability (Fig. 9.5) to get normalized values could be more complicated. For this system, it should be noted that ($S_o = 1-S_g-S_{wirr}$). Referring to Fig. 9.5, the relative permeability to gas ranges theoretically from 0.0 to 1.0 which corresponds to range of gas saturation from 0.0 to 1.0 as well. The practical range of the relative permeability to gas is from 0.0 to the maximum value of (K_{rg} @ $1-S_{org}-S_{wirr}$) which corresponds to gas saturation range from S_{gcr} to $1-S_{org}-S_{wirr}$. Considering this, the normalized gas saturation and the normalized relative permeability to gas can be calculated as;

$$S_g^* = (S_g - S_{gcr})/(1 - S_{org} - S_{wirr} - S_{gcr}) \tag{9.1.13}$$

$$K_{rg}^* = K_{rg}/(K_{rg}@1 - S_{org} - S_{wirr}) \tag{9.1.14}$$

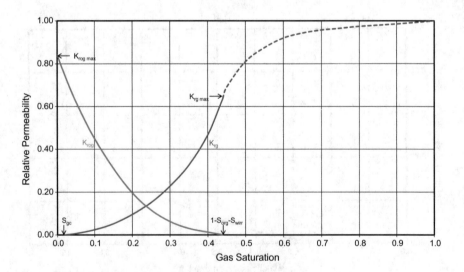

Fig. 9.5 Gas oil relative permeability ranges for normalization

Similarly, the relative permeability to oil ranges theoretically from 0.0 to 1.0 which corresponds to range of oil saturation from 0.0 to 1.0 as well. The practical range of relative permeability to oil is from 0.0 to the maximum value of (K_{ro} @ zero gas saturation) which corresponds to oil saturation range from S_{org} to $1-S_{wirr}$. Considering this, the normalized oil saturation and the normalized relative permeability to oil can be calculated as;

$$S_o^* = (S_o - S_{org})/(1 - S_{wirr} - S_{org}) \qquad (9.1.15)$$

$$K_{ro}^* = K_{ro}/(K_{ro} \text{ @ zero gas saturation}) \qquad (9.1.16)$$

Knowing that ($S_o = 1-S_g- S_{wirr}$), the Eq. 9.1.15 can be re-written as;

$$S_o^* = (1 - S_g - S_{wirr} - S_{org})/(1 - S_{wirr} - S_{org}) \qquad (9.1.17)$$

Note that in the case of ($S_{gcr} = 0.0$),

$$S_o^* = 1 - S_g^* \qquad (9.1.18)$$

While performing normalization and/or de-normalization procedure, it is advised to treat each phase independently.

In Eq. 9.1.4, Corey exponent has different values for different fluids in different rock types. Table 9.2 shows typical Corey exponents values for different fluid systems.

In all preceding discussions, water oil system and gas oil system were considered. These two phase systems are sufficient for many flow cases. The question arises; what would be the relative permeability to different phases in the three phase system? The **three phase relative permeability** needs to be considered where all the three phases (oil, gas and water) are mobile. This situation can arise in a reservoir under simultaneous gas and water drive. The importance of the three phase relative

Table 9.2 Typical corey exponents	Relative permeability curve	Wettability	Corey exponent
	K_{rw}	Water wet	5 to 8
		Intermediate wet	3 to 5
		Oil Wet	2 to 3
	K_{row}	Water wet	2 to 4
		Intermediate wet	3 to 6
		Oil wet	6 to 8
	K_{rg}	–	1 to 2
	K_{rog}	–	4 to 8

permeability is increased when studying enhanced oil recovery (EOR) applications (e.g.; the injection of gas in a watered-out reservoir).

Unfortunately, the estimation of the three phase relative permeability is a very difficult task. Since 1941, a lot of experimental work has been tried by many scholars to estimate the three phase relative permeability in laboratory. In addition to their difficulty, these trials ended with quite different results. This makes such results suspected and not conclusive. The other alternative to estimate the three phase relative permeability is the use of mathematical models. These models estimate the three phase relative permeability using the available two phase (water–oil system and gas-oil system) relative permeability information. In establishing such mathematical models, it is understood that the relative permeability to water is a function only of the water saturation. Similarly, the relative permeability to gas is a function only of the gas saturation. Relative permeability to oil is more complicated and can be a function of both water saturation and gas saturation.

$$K_{rw} = f(S_w) \tag{9.1.19}$$

$$K_{rg} = f(S_g) \tag{9.1.20}$$

$$K_{ro} = f(S_w, S_g) \tag{9.1.21}$$

Stone (1970) Model-1 and Stone (1973) Model-2 could be the most popular mathematical models to estimate the relative permeability to oil in three phase system. In addition to other mathematical models, the modified versions of Stone Model-1 and Stone Model-2 are built in Eclipse simulator.

As for Stone Model-1, the relative permeability to oil in the three phase system is estimated by;

$$K_{ro} = S_o^* \beta_w \beta_g \tag{9.7.22}$$

where,

$$S_o^* = (S_o - S_{om})/(1 - S_{wirr} - S_{om}) \tag{9.1.23}$$

$$\beta_w = K_{row}/(1 - S_w^*) \tag{9.1.24}$$

$$S_w^* = (S_w - S_{wirr})/(1 - S_{wirr} - S_{om}) \tag{9.1.25}$$

$$\beta_g = K_{rog}/(1 - S_g^*) \tag{9.1.26}$$

$$S_g^* = S_g/(1 - S_{wirr} - S_{om}) \tag{9.1.27}$$

Equation 9.1.22 can be re-written as;

$$K_{ro} = S_o^*\left[\left(K_{row}/(1 - S_W^*)\right)\right]\left[K_{rog}/(1 - S_g^*)\right] \qquad (9.1.28)$$

In the above, equations K_{ro} is the relative permeability to oil in the three phase system. S_o^*, S_w^* and S_g^* are normalized oil, water and gas saturation introduced by Stone to establish his model. S_{om} is defined as the minimum oil saturation in the system.

Aziz and Settari (1979) suggested a modification to Stone Model-1 to avoid calculating unrealistic values of K_{ro} (K_{ro} greater than 1.0). The modified Stone Model-1 calculates the three phase relative permeability to oil as;

$$K_{ro} = S_o^*\left[\left(K_{row}\,K_{rog}\right)/(K_{row}@S_{wirr})\right]/\left[\left(1 - S_w^*\right)\left(1 - S_g^*\right)\right] \qquad (9.1.29)$$

Choosing S_{om} value could be the main difficulty in using Stone Model-1 for the three phase relative permeability estimation. It is to be mentioned that S_{om} is different from the known S_{orw} (residual oil saturation in water–oil system) and S_{org} (residual oil saturation in gas-oil system). S_{om} was considered as constant in Stone Model-1. However, Fayers and Mathews (1984) suggested calculating different values or S_{om} using;

$$S_{om} = \alpha\,S_{orw} + (1 - \alpha)S_{org} \qquad (9.1.29)$$

where,

$$\alpha = 1 - \left[S_g/(1 - S_{wirr} - S_{org})\right] \qquad (9.1.30)$$

In the absence of other values, Eclipse simulator defaults S_{om} to the minimum residual oil saturation (minimum of S_{orw} and S_{org}).

Stone Model-2 was introduced to avoid the difficulty of choosing S_{om} value required for Stone Model-1. For this model, the relative permeability to oil in the three phase system is calculated by:

$$K_{ro} = (K_{row} + K_{rw})(K_{rog} + K_{rg}) - (K_{rw} + K_{rg}) \qquad (9.1.31)$$

Similar to Stone Model-1, a modified version of Stone Model-2 was suggested by Aziz and Settari;

$$K_{ro} = (K_{row}@S_{wirr})[\{\, K_{row}/(K_{row}@S_{wirr}) + K_{rw}\}\{\, K_{rog}/(K_{row}@S_{wirr}) + K_{rg}\}$$
$$- (K_{rw} + K_{rg})] \qquad (9.1.32)$$

It is to be noted that negative values of K_{ro} may be produced by using Stone Model-2 or its modified version. For such case, Eclipse automatically changes the calculated value of K_{ro} to zero.

The difficulty and non-conclusive estimation of the three phase relative permeability led Baker (1988) to conclude: 'There are still many problems to be solved'. He subsequently suggested that the three phase relative permeability to oil would be estimated by simple linear interpolation between the water–oil and gas-oil relative permeability data;

$$K_{ro} = \{K_{rog}(S_g - S_{gcr}) + K_{row}(S_w - S_{wirr})\}/\{(S_g - S_{gcr}) + (S_w - S_{wirr})\}$$

(9.1.33)

Or more simply;

$$K_{ro} = (K_{rog}S_g + K_{row}S_w)/(S_g + S_w)$$

(9.1.34)

In reservoir simulation work, it is recommended to simply use the default simulator built-in three phase relative permeability model to avoid the difficulty and uncertainty of additional data requirement.

Exercises

(e)   Do the necessary adjustment for the water–oil relative permeability data in Table 9.1.

(f)   Calculate the normalized relative permeability for the data of the previous exercise.

10   Prepare a full set of relative permeability curves for shore face sand;

Permeability 200 md

Porosity   20 %

11   The sand of the previous exercise has been subject to gas injection.
At one point of time the water, oil and gas saturations were 60, 30 and 10% respectively.

Estimate the three phase permeability to oil at this point of time using Stone 1, Stone 2 and Baker models

9.2   Rock Typing and Rock Groups

Rock typing is the process of splitting the reservoir rock into different groups according to the change of some or all rock properties. The rock typing is normally done by geoscientists and come out as one output of the reservoir static model work. It is believed that parts of the reservoir with the same rock type would share the same range of rock properties. The poro-perm plot of Fig. 9.6 illustrates data of five different rock types in a hydrocarbon reservoir. It can be simply recognized that each rock type has a distinct range of permeability. However, this ideal case is seldom existing.

In dynamic model work, the reservoir engineer normally assigns different saturation functions (relative permeability/ capillary pressure) sets to different parts of the reservoir according to ranges of Porosity (Φ), Permeability (K), Rock Quality Index (RQI) or Flow Zone Index (FZI). The rock quality index (RQI) is calculated by:

$$RQI = Sqrt(K/\Phi) \qquad (9.2.1)$$

The Flow Zone Index is calculated by:

$$FZI = RQI/\Phi_z \qquad (9.2.2)$$

where,

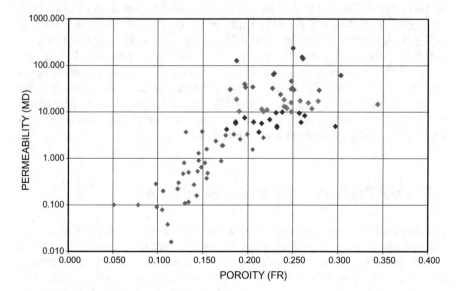

Fig. 9.6 Permeability versus porosity (Popo-Perm)

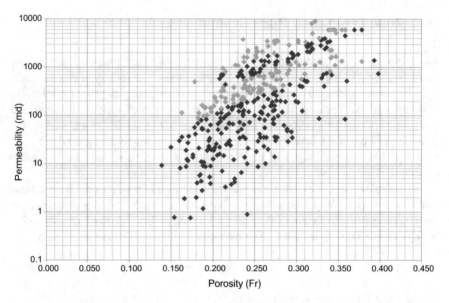

Fig. 9.7 Popo-Perm relation

$$\Phi_z = \text{Normalized Porosity} = \Phi/(1 - \Phi) \qquad (9.2.3)$$

Rock quality index (RQI) could be the most popular parameter that being used for this purpose. Due to considering more descriptive information, the static model rock types' Φ, K, RQI and FZI ranges may overlap each other. The poro-perm plot of Fig. 9.7 illustrates a case where no distinct range of K can be recognized for each rock type. Consequently, no distinct range of a controlling parameter (K, RQI or FZI) can be expected for each rock type. For such a case, the reservoir engineer should consider sub-grouping within each static model rock type according to one or more of the mentioned controlling parameters. Figures 9.8 illustrates three static model rock types whose ranges of permeability overlap each other. Figures 9.9 and 9.10 illustrate two different schemes for subgrouping.

9.3   Rock Properties in Reservoir Simulation

Building a representative reservoir simulation model is an essential part in most of the field development plan (FDP) work. The task of building a reservoir simulation models needs to collect, validate and properly formulate different reservoir data sets. Basic rock properties (porosity and permeability) distribution is supplied as a product of the static model work. As such the RE primarily has no direct role in distributing these properties. However, he/she should keep close communication with other team members in order to be aware of any issues relating to population of these parameters. On the other hand, he/she should provide any reservoir engineering

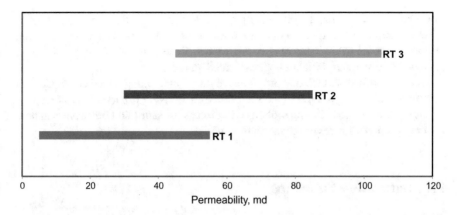

Fig. 9.8 Rock types and rock grouping

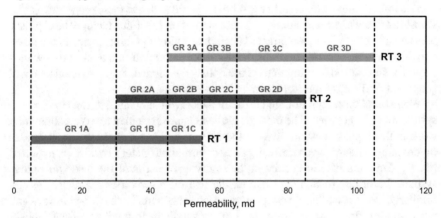

Fig. 9.9 Rock types and rock grouping

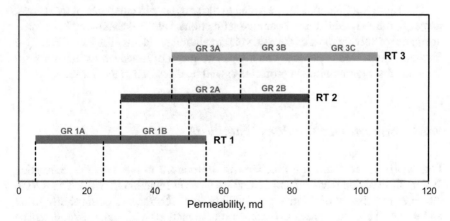

Fig. 9.10 Rock types and rock grouping

information that can help in the proper population of these parameters. Additionally, he/she is supposed to take a good share in the procedure of up scaling from the fine static model to the coarser dynamic model. Reservoir rock compressibility is provided as a standard or laboratory evaluated value/s.

Other important part of rock properties required in building the reservoir simulation model is the preparation of the saturation functions. This part is the full responsibility of the RE. The coming sections will discuss the saturation functions and how they are prepared for reservoir simulation model.

9.4 Saturation Functions

Saturation Functions are the reservoir rock properties that change mainly as functions to fluid saturation change. Some scholars prefer to define these properties as **The Combined Rock Fluid Properties.** These saturation functions include the capillary pressure and relative permeability as function of fluid saturation change in the reservoir. The coming sections describe details about saturation functions and how they should be introduced to simulator. As previously mentioned, our discussion here will consider ECLIPSE as the simulator.

When preparing saturation functions for a reservoir simulator, the reservoir phase system **must** be honored. The most generalized phase system in reservoir simulation work is the 3-phase system. This is the system where the three reservoir phases (hydrocarbon liquid, hydrocarbon gas and formation water) coexist or expected (during depletion life) to coexist in the reservoir or part/s of the reservoir porous system. According to this explanation, the live oil reservoir is a 3-phase system. Similarly, the gas condensate reservoir is a 3-phase system. For such 3-phase system, the saturation functions are introduced to the simulator in the form of the 2-phase sets, namely the oil–water and the gas-oil systems.

The saturation functions are provided to the simulator in either one of two forms; namely Family-1 and Family-2 Saturation functions. Family-1 Saturation Function is adequate for solving most of the reservoir simulation problems. The use of Family-2 Saturation Function is generally an alternative method. However, its use is a must for some reservoir simulation problems as will be discussed in a latter section.

9.4.1 Saturation Functions (Family-1)

For Family-1, the saturation functions are introduced to the model as Saturation Water Oil Function (SWOF) and Saturation Gas Oil Function (SGOF). The SWOF tabulates the change of relative permeability to water, relative permeability to oil and water oil capillary pressure (or water oil J-Function) with water saturation. The

SGOF tabulates the change of relative permeability to gas, relative permeability to oil and the gas oil capillary pressure (or gas oil J-Function) with gas saturation.

Tables 9.3 and 9.4 are typical saturation functions' tables for both oil–water system (SWOF) and gas–oil system (SGOF). The same tabulated functions are shown graphically in Figs. 9.11, 9.12, 9.13 and 9.14. As previously explained while discussing about relative permeability, it is important to understand that the water–oil system (SWOF) means (water + oil) and gas-oil system (SGOF) means (gas + oil + irreducible water).

The SGOF tabulated function can be **arbitrarily** replaced by SLGOF (gas–liquid system) function. For this case, the gas-oil system relative permeability and capillary pressure are tabulated as a function of liquid (oil + irreducible water) saturation. Table 9.5 shows the SLGOF function that may replace the given SGOF function (Table 9.4). Refer also to Figs. 9.15 and 9.16 for graphical illustration of the SLGOF functions.

9.5   Saturation Functions (Family-2)

For Family-2, the saturation functions are introduced to the model as Saturation Water Function (SWFN), Saturation Gas Function (SGFN) and Saturation Oil Function (SOF3). The SWFN represents the change of relative permeability to water and water oil capillary pressure with water saturation. The SGFN represents the change of relative permeability to gas and the gas oil capillary pressure with gas saturation. The Oil Saturation Function (SOF3) represents the change of relative permeability to oil in both water–oil and gas-oil systems. Tables 9.6, 9.7 and 9.8 are typical saturation functions' tables considering Family-2 option. These tables tabulate the same data of the previous section (Family-1) in the alternative way (Family-2). Naturally, the graphical presentation of the two methods is the same.

9.6   Drainage and Imbibition Saturation Functions

Examining the tables in previous sections, it can be noticed that each table contains two sub-tables. The two sub-tables are almost similar. These are recognized by the simulator as the Drainage and Imbibition Saturation Tables. The standard definition of drainage and imbibition is not necessarily applicable here. The simulator uses the drainage table/s in the reservoir model initialization step. This means that the capillary pressure/J-Function is the only significant information for this step. For all steps following initialization, the simulator uses the imbibition table/s. However, the simulator will shuttle between the two groups (imbibition and drainage) if the **hysteresis** option is activated. If the imbibition table/s group is not entered, the simulator will use the drainage group for all steps including initialization.

Table 9.3 SWOF

S_w	K_{rw}	K_{row}	P_c		
0.260	0.000000	0.841000	45.0000	Min S_w (S_{wirr})	Max K_{row}
0.267	0.000000	0.780000	43.5200	Critical S_w (S_{wcr})	
0.300	0.008301	0.570000	37.4633		
0.350	0.012484	0.370000	32.7597		
0.400	0.018309	0.230000	29.4844		
0.450	0.026602	0.130000	27.2550		
0.500	0.037638	0.065000	25.0000		
0.550	0.052884	0.028000	23.0000		
0.600	0.071386	0.005000	21.3091		
0.650	0.100064	0.000000	19.5000	$S_w = 1 - S_{orw}$	Max K_{rw}
0.700	0.145000	0.000000	17.7000		
0.750	0.210000	0.000000	15.5000		
0.800	0.310000	0.000000	12.9611		
0.850	0.430000	0.000000	10.3823		
0.900	0.575000	0.000000	7.5385		
0.950	0.750000	0.000000	4.1046		
0.975	0.860000	0.000000	1.0000		
1.000	1.000000	0.000000	0.0000	Max S_w (100%)	
S_w	K_{rw}	K_{row}	P_c		
0.260	0.000000	0.841000	45.0000	Min S_w (S_{wirr})	Max K_{row}
0.267	0.000000	0.780000	30.0000	Critical S_w (S_{wcr})	
0.300	0.008301	0.570000	19.0000		
0.350	0.012484	0.370000	11.4297		
0.400	0.018309	0.230000	8.6241		
0.450	0.026602	0.130000	6.6317		
0.500	0.037638	0.065000	5.0942		
0.550	0.052884	0.028000	4.0656		
0.600	0.071386	0.005000	2.5000		
0.650	0.100064	0.000000	0.0000	$S_w = 1 - S_{orw}$	Max K_{rw}
0.700	0.145000	0.000000	0.0000		
0.750	0.210000	0.000000	0.0000		
0.800	0.310000	0.000000	0.0000		
0.850	0.430000	0.000000	0.0000		
0.900	0.575000	0.000000	0.0000		
0.950	0.750000	0.000000	0.0000		
0.975	0.860000	0.000000	0.0000		
1.000	1.000000	0.000000	0.0000	Max S_w (100%)	

Table 9.4 SGOF

S_g	K_{rg}	K_{rog}	P_c		
0.000	0.000000	0.841000	0.0000	Min S_g	Max K_{rog}
0.020	0.000000	0.748674	0.4926	Critical S_g (S_{gcr})	
0.050	0.006145	0.623689	1.2314		
0.100	0.024128	0.449028	2.2616		
0.150	0.052081	0.310298	3.1147		
0.200	0.094997	0.196275	3.8883		
0.250	0.152222	0.114033	4.5216		
0.300	0.228387	0.061589	5.1987		
0.350	0.325759	0.029673	5.6720		
0.400	0.467025	0.012494	6.3927		
0.433	0.637678	0.000000	6.7384	$S_g = 1 - S_{org} - S_{wirr}$	Max K_{rg}
0.450	0.690000	0.000000	6.9430		
0.500	0.810000	0.000000	7.5761		
0.550	0.875000	0.000000	8.1765		
0.600	0.920000	0.000000	8.8453		
0.650	0.943000	0.000000	9.8279		
0.700	0.957000	0.000000	11.2390		
0.740	0.963000	0.000000	13.0560	Max $S_g = 1 - S_{wirr}$	
0.000	0.000000	0.841000	0.0000	Min S_g	Max K_{rog}
0.020	0.000000	0.748674	0.4926	Critical S_g (S_{gcr})	
0.050	0.006145	0.623689	1.2314		
0.100	0.024128	0.449028	2.2616		
0.150	0.052081	0.310298	3.1147		
0.200	0.094997	0.196275	3.8883		
0.250	0.152222	0.114033	4.5216		
0.300	0.228387	0.061589	5.1987		
0.350	0.325759	0.029673	5.6720		
0.400	0.467025	0.012494	6.3927		
0.433	0.637678	0.000000	6.7384	$S_g = 1 - S_{org} - S_{wirr}$ Max K_{rg}	
0.450	0.690000	0.000000	6.9430		
0.500	0.810000	0.000000	7.5761		
0.550	0.875000	0.000000	8.1765		
0.600	0.920000	0.000000	8.8453		
0.650	0.943000	0.000000	9.8279		
0.700	0.957000	0.000000	11.2390		
0.740	0.963000	0.000000	13.0560	Max $S_g = 1 - S_{wirr}$	

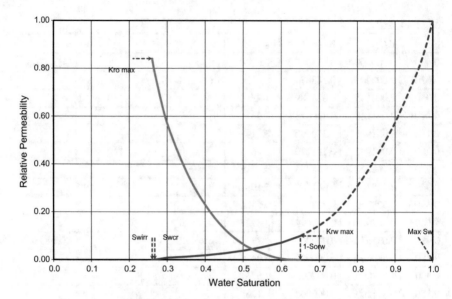

Fig. 9.11 Water oil relative permeability

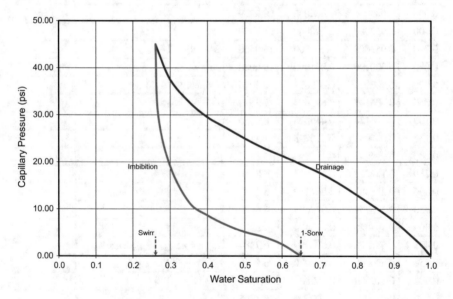

Fig. 9.12 Water oil capillary pressure

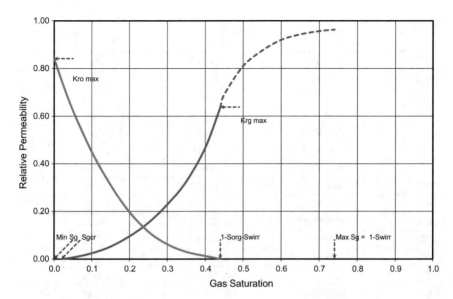

Fig. 9.13   Gas oil relative permeability

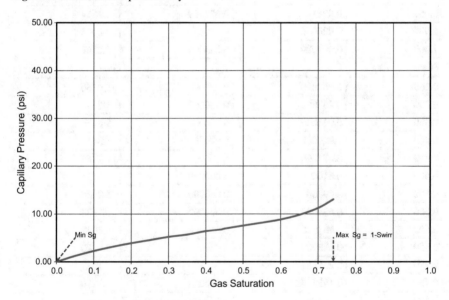

Fig. 9.14   Gas oil capillary pressure

Table 9.5   SLGOF

S_l	K_{rg}	K_{rog}	P_c
0.260	0.963000	0.000000	13.0560
0.300	0.957000	0.000000	11.2390
0.350	0.943000	0.000000	9.8279
0.400	0.920000	0.000000	8.8453
0.450	0.875000	0.000000	8.1765
0.500	0.810000	0.000000	7.5761
0.550	0.690000	0.000000	6.9430
0.560	0.637678	0.000000	6.7384
0.600	0.467025	0.012494	6.3927
0.650	0.325759	0.029673	5.6720
0.700	0.228387	0.061589	5.1987
0.750	0.152222	0.114033	4.5216
0.800	0.094997	0.196275	3.8883
0.850	0.052081	0.310298	3.1147
0.900	0.024128	0.449028	2.2616
0.950	0.006145	0.623689	1.2314
0.980	0.000000	0.748674	0.4926
1.000	0.000000	0.841000	0.0000
S_l	K_{rg}	K_{rog}	P_c
0.260	0.963000	0.000000	13.0560
0.300	0.957000	0.000000	11.2390
0.350	0.943000	0.000000	9.8279
0.400	0.920000	0.000000	8.8453
0.450	0.875000	0.000000	8.1765
0.500	0.810000	0.000000	7.5761
0.550	0.690000	0.000000	6.9430
0.560	0.637678	0.000000	6.7384
0.600	0.467025	0.012494	6.3927
0.650	0.325759	0.029673	5.6720
0.700	0.228387	0.061589	5.1987
0.750	0.152222	0.114033	4.5216
0.800	0.094997	0.196275	3.8883
0.850	0.052081	0.310298	3.1147
0.900	0.024128	0.449028	2.2616
0.950	0.006145	0.623689	1.2314
0.980	0.000000	0.748674	0.4926
1.000	0.000000	0.841000	0.0000

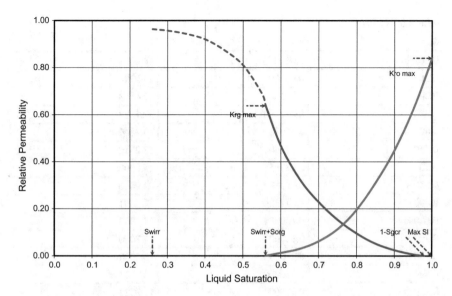

Fig. 9.15   Gas oil relative permeability

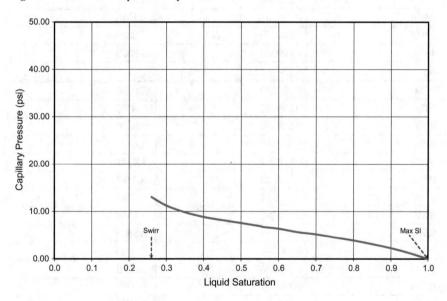

Fig. 9.16   Gas oil capillary pressure

Table 9.6 SWFN

S_w	K_{rw}	P_c
0.260	0.000000	45.0000
0.267	0.000000	43.5200
0.300	0.008301	37.4633
0.350	0.012484	32.7597
0.400	0.018309	29.4844
0.450	0.026602	27.2550
0.500	0.037638	25.0000
0.550	0.052884	23.0000
0.600	0.071386	21.3091
0.650	0.100064	19.5000
0.700	0.145000	17.7000
0.750	0.210000	15.5000
0.800	0.310000	12.9611
0.850	0.430000	10.3823
0.900	0.575000	7.5385
0.950	0.750000	4.1046
0.975	0.860000	1.0000
1.000	1.000000	0.0000
S_w	K_{rw}	P_c
0.260	0.000000	45.0000
0.267	0.000000	30.0000
0.300	0.008301	19.0000
0.350	0.012484	11.4297
0.400	0.018309	8.6241
0.450	0.026602	6.6317
0.500	0.037638	5.0942
0.550	0.052884	4.0656
0.600	0.071386	2.5000
0.650	0.100064	0.0000
0.700	0.145000	0.0000
0.750	0.210000	0.0000
0.800	0.310000	0.0000
0.850	0.430000	0.0000
0.900	0.575000	0.0000
0.950	0.750000	0.0000
0.975	0.860000	0.0000
1.000	1.000000	0.0000

Table 9.7 SGFN

S_g	K_{rg}	P_c
0.000	0.000000	0.0000
0.020	0.000000	0.4926
0.050	0.006145	1.2314
0.100	0.024128	2.2616
0.150	0.052081	3.1147
0.200	0.094997	3.8883
0.250	0.152222	4.5216
0.300	0.228387	5.1987
0.350	0.325759	5.6720
0.400	0.467025	6.3927
0.433	0.637678	6.7384
0.450	0.690000	6.9430
0.500	0.810000	7.5761
0.550	0.875000	8.1765
0.600	0.920000	8.8453
0.650	0.943000	9.8279
0.700	0.957000	11.2390
0.740	0.963000	13.0560
S_g	K_{rg}	P_c
0.000	0.000000	0.0000
0.020	0.000000	0.4926
0.050	0.006145	1.2314
0.100	0.024128	2.2616
0.150	0.052081	3.1147
0.200	0.094997	3.8883
0.250	0.152222	4.5216
0.300	0.228387	5.1987
0.350	0.325759	5.6720
0.400	0.467025	6.3927
0.433	0.637678	6.7384
0.450	0.690000	6.9430
0.500	0.810000	7.5761
0.550	0.875000	8.1765
0.600	0.920000	8.8453
0.650	0.943000	9.8279
0.700	0.957000	11.2390
0.740	0.963000	13.0560

Table 9.8 SOF3

S_o	K_{row}	K_{rog}
0.000	0.000000	0.000000
0.300	1*	0.000000
0.340	1*	0.012494
0.350	0.000000	1*
0.390	1*	0.029673
0.400	0.005000	1*
0.440	1*	0.061589
0.450	0.028000	1*
0.490	1*	0.114033
0.500	0.065000	1*
0.540	1*	0.196275
0.550	0.130000	1*
0.590	1*	0.310298
0.600	0.230000	1*
0.640	1*	0.449028
0.650	0.370000	1*
0.690	1*	0.623689
0.700	0.570000	1*
0.720	1*	0.748674
0.733	0.780000	1*
0.740	0.841000	0.841000
S_o	K_{row}	K_{rog}
0.000	0.000000	0.000000
0.300	1*	0.000000
0.340	1*	0.012494
0.350	0.000000	1*
0.390	1*	0.029673
0.400	0.005000	1*
0.440	1*	0.061589
0.450	0.028000	1*
0.490	1*	0.114033
0.500	0.065000	1*
0.540	1*	0.196275
0.550	0.130000	1*
0.590	1*	0.310298
0.600	0.230000	1*
0.640	1*	0.449028

(continued)

Table 9.8 (continued)

S_o	K_{row}	K_{rog}
0.650	0.370000	1*
0.690	1*	0.623689
0.700	0.570000	1*
0.720	1*	0.748674
0.733	0.780000	1*
0.740	0.841000	0.841000

9.7 Saturation Functions for Water Gas System (Gas Reservoir Problems)

As previously mentioned, phase system of the reservoir should be honored when preparing the saturation functions for reservoir simulation model. A gas reservoir can be one of three types; namely dry gas, wet gas or gas condensate reservoir. For dry gas reservoir, liquid hydrocarbon is not expected to drop out inside the reservoir nor in the tabular facilities for the whole reservoir life. For wet gas reservoir, liquid hydrocarbon is not likely to drop out inside the reservoir but is expected to drop out in the tabular facilities. For gas condensate reservoir, liquid hydrocarbon is expected to drop out inside the reservoir. From this description, the wet gas and gas condensate reservoirs should be considered as 3-phase systems. The saturation tables structure discussed in the previous sections are applicable for such reservoirs. It is well known that these reservoirs initially contain only hydrocarbon gas (wet gas or gas condensate) and water phases. The liquid hydrocarbon phase (condensate) is expected to drop out later during depletion. To allow the simulator to properly initialize the model, the water–oil capillary pressure/J-function in the SWOF tables must be replaced by the water–gas capillary pressure/J-function. The gas oil capillary pressure entries in SGOF table are to be set to zero.

Dry Gas Reservoir is a good example for 2-phase reservoir system. In a dry gas reservoir, no liquid hydrocarbon is expected to exist in the reservoir pore space at any time of the reservoir life. It is also an example where the use of Family-2 Saturation Functions method is compulsory. SWFN represents the change of relative permeability to water and water–gas capillary pressure (or water–gas J-function) with water saturation (Table 9.9). SGFN represents the change of relative permeability to gas with gas saturation. For this tabular function, all entries of the third column reserved for gas-oil capillary pressure must be zero (Table 9.10). Figures 9.17 and 9.18 are graphical presentation for the two functions.

Another alternative for dry gas reservoir saturation functions is to use SGWFN function instead of the two functions SWFN and SGFN. The SGWFN function tabulates relative permeability to gas, relative permeability to water and water gas capillary pressure (or J-Function) changes with gas saturation. Table 9.11 shows the alternative SGWFN function that would replace the SWFN and SGFN functions of the previous tables.

Table 9.9   SWFN

S_w	K_{rw}	J (W-G)	
0.2500	0.000000	8.91689	Min S_w (S_{wirr})/Critical S_w (S_{wcr})
0.2775	0.000002	5.58906	
0.3050	0.000030	3.66142	
0.3325	0.000152	2.48788	
0.3600	0.000480	1.74322	
0.3875	0.001172	1.25386	
0.4150	0.002430	0.92250	
0.4425	0.004502	0.69221	
0.4700	0.007680	0.52848	
0.4975	0.012302	0.40972	
0.5250	0.018750	0.32203	
0.5525	0.027452	0.25624	
0.5800	0.038880	0.20617	
0.6075	0.053552	0.16756	
0.6350	0.072030	0.13743	
0.6625	0.094922	0.11368	
0.6900	0.122880	0.09476	
0.7175	0.156602	0.07955	
0.7450	0.196830	0.06722	
0.7725	0.244352	0.05716	
0.8000	0.300000	0.04887	$S_w = 1 - S_{gr}$
0.8500	0.500000	0.03726	
0.9000	0.750000	0.02885	
0.9500	0.900000	0.02265	
1.0000	1.000000	0.00000	Max S_w (1.0)

Examining the plots of Figs. 9.17 and 9.18, it is possible to conclude that the saturation functions for water–gas system is quiet similar to the one for water–oil system. The residual gas saturation/trapped gas saturation (S_{gr}/S_{gt}) in the water–gas system replaces the residual oil saturation (S_{orw}) in the water–oil system. Also, the maximum relative permeability to gas ($K_{rg\ max}$ or K_{rg} @ S_{wirr}) in the water–gas system replaces the maximum relative permeability to oil ($K_{ro\ max}$ or K_{row} @ S_{wirr}) in the water–oil system.

Table 9.10 SGFN

S_g	K_{rg}	J (G-O)	
0.0000	0.00000	0.00000	Min S_g (0.0)
0.0500	0.00000	0.00000	
0.1000	0.00000	0.00000	
0.1500	0.00000	0.00000	
0.2000	0.00000	0.00000	Residual Gas Saturation (S_{gr})
0.2275	0.00238	0.00000	
0.2550	0.00950	0.00000	
0.2825	0.02138	0.00000	
0.3100	0.03800	0.00000	
0.3375	0.05937	0.00000	
0.3650	0.08550	0.00000	
0.3925	0.11638	0.00000	
0.4200	0.15200	0.00000	
0.4475	0.19238	0.00000	
0.4750	0.23750	0.00000	
0.5025	0.28738	0.00000	
0.5300	0.34200	0.00000	
0.5575	0.40138	0.00000	
0.5850	0.46550	0.00000	
0.6125	0.53438	0.00000	
0.6400	0.60800	0.00000	
0.6675	0.68638	0.00000	
0.6950	0.76950	0.00000	
0.7225	0.85738	0.00000	
0.7500	0.95000	0.00000	Max Gas Saturation ($1-S_{wirr}$)

Exercise

12   The following two tables show the SWFN and SGFN functions for a Water-Gas system reservoir.

Prepare the alternative SGWFN function for the same reservoir.

SWFN	S_w	K_{rw}	J(wat-gas)	SGFN	S_g	K_{rg}	J(gas-oil)
$S_{wirr} = S_{wcr}$	0.2500	0.000000	8.8169	Min S_g	0.0000	0.00000	0.00
	0.2775	0.000002	5.5891		0.0500	0.00000	0.00
	0.3050	0.000030	3.6614		0.1000	0.00000	0.00
	0.3875	0.001172	1.2539		0.1500	0.00000	0.00
	0.4700	0.007680	0.5285	S_{gr}	0.2000	0.00000	0.00

(continued)

(continued)

SWFN	S_w	K_{rw}	J(wat-gas)	SGFN	S_g	K_{rg}	J(gas-oil)
	0.5525	0.027450	0.2542		0.2275	0.00250	0.00
	0.6350	0.070230	0.1374		0.2825	0.02138	0.00
	0.7175	0.156600	0.0796		0.3650	0.08550	0.00
	0.7725	0.244352	0.0572		0.4475	0.19238	0.00
$1-S_{gr}$	0.8000	0.300000	0.0489		0.5300	0.34200	0.00
	0.8500	0.500000	0.0373		0.6125	0.53438	0.00
	0.9000	0.750000	0.0289		0.6950	0.76950	0.00
	0.9500	0.950000	0.0227		0.7225	0.85738	0.00
Max S_w	1.0000	1.000000	0.0000	Max S_{gr}	0.7500	0.95000	0.00

9.8   Saturation Functions' End Points

Saturation Functions' End Points are the entries in the Saturation Functions' Tables that the simulator interprets for specific meaning. As an example, these entries are highlighted in different colors in the previously tabulated saturation functions (Family-1, SWOF and SGOF, Tables 9.3 and 9.4). The physical definitions of these end points are well known and have nothing special. However, the way they are recognized and interpreted by simulator should be honored when building related reservoir simulation. Erroneous end points or not honoring the way they are interpreted by the simulator will lead to mis-interpretation and will affect the simulation results. The careful step by step evaluation of the end points before being introduced to the simulator is always preferred. If some built-in software is used to prepare the saturation functions, the validity of the resulted end points should be carefully checked. It is believed that visual examination of the resulted saturation functions curves is not enough to assure the validity of the end points.

In the following sections, each saturation function end point will be discussed in some details. In the discussion, the reference is made to Family-1 (SWOF and SGOF) saturation functions (Tables 9.3 and 9.4). Sections 9.8.1 to 9.8.4 discuss the water oil system end points while Sects. 9.8.5 to 9.8.8 deal with the gas oil system end points. In spite of discussing these end points considering Family-1 form of saturation functions, it is not difficult to extend the understanding to other forms of saturation functions.

9.8.1   Mimum Water Saturation (Eclipse: Swl)

This is referred physically to the irreducible water saturation (S_{wirr}). It is the minimum water saturation in the reservoir system and it is the first saturation entry in SWOF function table (Table 9.3). Initial water saturation will be set to this value for all points in the reservoir that lie above the top of water oil transition zone. In other

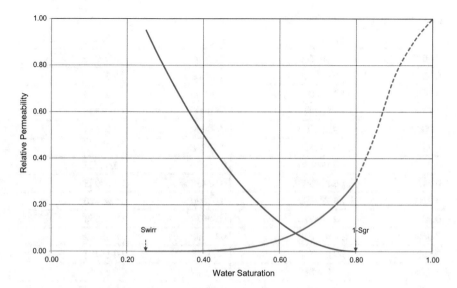

Fig. 9.17 Water gas relative permeability

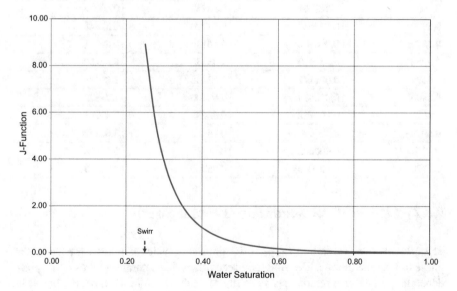

Fig. 9.18 Gas water J-function

words, initial water saturation will be set to this value for all points with initial water oil capillary pressure higher than the value corresponding to this saturation (first P_c entry in SWOF function). At this water saturation value, the relative permeability to water (K_{rw}) is equal to zero and the relative permeability to oil (K_{row}) is at its maximum (K_{row} @ S_{wirr}).

Table 9.11 SGWFN

S_g	K_{rw}	K_{rg}	J (W-G)
0.0000	1.000000	0.00000	0.00000
0.0500	0.900000	0.00000	0.02265
0.1000	0.750000	0.00000	0.02885
0.1500	0.500000	0.00000	0.03726
0.2000	0.300000	0.00000	0.04887
0.2275	0.244352	0.00238	0.05716
0.2550	0.196830	0.00950	0.06722
0.2825	0.156602	0.02138	0.07955
0.3100	0.122880	0.03800	0.09476
0.3375	0.094922	0.05937	0.11368
0.3650	0.072030	0.08550	0.13743
0.3925	0.053552	0.11638	0.16756
0.4200	0.038880	0.15200	0.20617
0.4475	0.027452	0.19238	0.25624
0.4750	0.018750	0.23750	0.32203
0.5025	0.012302	0.28738	0.40972
0.5300	0.007680	0.34200	0.52848
0.5575	0.004502	0.40138	0.69221
0.5850	0.002430	0.46550	0.92250
0.6125	0.001172	0.53438	1.25386
0.6400	0.000480	0.60800	1.74322
0.6675	0.000152	0.68638	2.48788
0.6950	0.000030	0.76950	3.66142
0.7225	0.000002	0.85738	5.58906
0.7500	0.000000	0.95000	8.91689

9.8.2 *Critical Water Saturation (Eclipse: Swcr)*

The critical water saturation (S_{wcr}) is the minimum water saturation for the water phase to become mobile. The simulator recognizes this saturation by its correspondence to the last zero relative permeability to water (K_{rw}) in the SWOF table. It is normal that the irreducible water saturation and the critical water are equal. For this case, one entry (S_{wirr}) in the SWOF is adequate to present both saturations.

9.8.3 Water Saturation at Residual Oil (Eclipse: 1−Sowcr)

At this water saturation, the oil saturation reaches its residual value (S_{orw}) in the water oil system. The simulator recognizes this saturation by its correspondence to the first zero relative permeability to oil (K_{row}) and consequently sets the value of the residual oil saturation (S_{orw}). At this saturation, the relative permeability to water (K_{rw}) reaches its maximum value as it can be estimated by laboratory experiment (K_{rw} @ S_{orw}). Naturally, the oil phase becomes immobile at this saturation.

9.8.4 Maximum Water Saturation (Eclipse: Swu)

This is the maximum water saturation ($S_{w\,max}$) in the reservoir system and it is the last entry in the SWOF tabulated function. The simulator will set this water saturation value to all points at the level of the given oil water contact (OWC) and deeper. Normally this value is 1.0 (100% water saturation). However, values less than 1.0 can be entered for some rare cases.

9.8.5 Minimum Gas Saturation (Eclipse: Sgl)

Minimum gas saturation ($S_{g\,min}$) is the first gas saturation entry in the SGOF table and it must be zero. Practically, it is the minimum possible gas saturation in the reservoir system. Any point in the reservoir that lies deeper than the given gas oil contact (GOC) will be initially set to this gas saturation value ($S_g = 0.0$). At this saturation value, relative permeability to gas (K_{rg}) is zero while the relative permeability to oil (K_{rog}) is at its maximum value. It is important to notice that the maximum K_{rog} (K_{rog} @ $S_{g\,min}$) is equal to the maximum K_{row} (K_{row} @ S_{wirr}). This is natural because for both cases the system contains only oil and irreducible water.

9.8.6 Critical Gas Saturation (Eclipse: Sgcr)

The critical gas saturation (S_{gcr}) is the minimum gas saturation for the gas phase to become mobile. The simulator recognizes this saturation by its correspondence to the last zero relative permeability to gas (K_{rg}) in the SGOF table. Normally, critical gas saturation is very small and can be considered as equal to zero. For zero critical gas saturation, one entry is adequate for both minimum and critical gas saturations in the SGOF table.

9.8.7 Gas Saturation at Residul Oil Saturation (Eclipse: 1—Sogcr—Swl)

At this gas saturation, the oil saturation reaches its residual value (S_{org}) in the gas oil system. This saturation is also defined as the gas saturation at residual liquid saturation (S_{lr}). The simulator recognizes this value by its correspondence to the first zero relative permeability to oil (K_{rog}) in the SGOF table. It consequently sets the residual oil saturation value (S_{org}) for the system. At this saturation, the relative permeability to gas (K_{rg}) reaches its maximum value as it can be estimated in laboratory (K_{rg} @ $S_{lr} = S_{org} + S_{wirr}$) by gas displacing oil experiment. It is clear that the oil phase becomes immobile at this saturation.

9.8.8 Maximum Gas Saturation (Eclipse: Sgu)

This is the maximum gas saturation ($S_{g\,max}$) in the reservoir system and it is the last entry in the SGOF tabulated function. The simulator will set this gas saturation value to all points shallower than the top of the gas oil transition zone. It is understood that this value should be equal to ($1-S_{wirr.}$). For this gas saturation, the relative permeability to gas (K_{rg}) is higher than the value explained in the previous section. Normally this relative permeability value is not available from laboratory experiment. It can be estimated by interpolation between the two values (K_{rg} @ S_{lr}) and ($K_{rg} = 1.0$ @ theoretical $S_g = 1.0$).

9.8.9 Violating End Points Rules (What Could Be the Effect?)

In this section, we are going to test the effect of violating some of the previously explained end points' rules on reservoir simulation model initialization (Fawakhiri et al., 1989). To clarify on this matter, let us consider the simple reservoir one direction model in Fig. 9.19. Let us also consider initializing this model using saturation Table 9.3 (SWOF) and 9.4 (SGOF) in Sect. 9.4.1. These tables fully honor Eclipse End Points' Rules. The initialization results are summarized in Table 9.12. These results are the ones that will be used as a basis of comparison. Refer also to Fig. 9.20 which shows different phases saturation with depth at initialization.

Let us now make two modifications to the SWOF (water oil system) saturation function table:

- Add one entry ($S_w = 0.0$) at the table start.
- Remove all entries after ($S_w = 1-S_{orw}$).

By these modifications the SWOF table would look as seen in Table 9.13

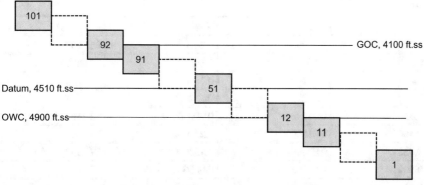

* One Dimension Model 101 * 1 * 1 Blocks
* Equal Blocks of 200 * 200 * 20 ft
* OWC @ 4900 ft.ss, GOC 4100 ft.ss
* Porosity 20 %
* Permeability 100 md

Fig. 9.19 Different phases saturation with depth at initialization

Table 9.12 Prepare the alternative SGWFN function for the same reservoir	Fluid	Initial volume in place	Initial saturation range
	Oil	1,114,722 stb	0.00–0.74
	Water	1,241,147 stb	0.26–1.00
	Free gas	190,050 Mscf	0.00–0.74

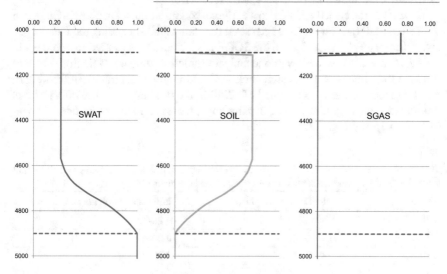

Fig. 9.20 Initial saturation profiles, honoring end point rules

Table 9.13 SWOF

S_w	K_{rw}	K_{row}	P_c
0.000	0.000000	0.841000	45.0000
0.260	0.000000	0.841000	45.0000
0.267	0.000000	0.780000	43.5200
0.300	0.008301	0.570000	37.4633
0.350	0.012484	0.370000	32.7597
0.400	0.018309	0.230000	29.4844
0.450	0.026602	0.130000	27.2550
0.500	0.037638	0.065000	25.0000
0.550	0.052884	0.028000	23.0000
0.600	0.071386	0.005000	21.3091
0.650	0.100064	0.000000	19.5000

The first look to the modified table does not indicate any technical problem. Thinking the same way as the simulator, we can conclude that the minimum and maximum water saturations in the system will be recognized as 0.0 and 0.65 respectively. The value of $S_w = 0.0$ will be set to all points above water–oil transition zone. The value of $S_w = 0.65$ will be set to all points deeper than the given OWC. The effect of this on evaluating the initial volumes in place and the initial saturation distribution are shown in Table 9.14 and Fig. 9.21.

Now, let us put back the SWOF to its original condition and try to play similar game with the SGOF (gas oil system) saturation function table. By ending the table at gas saturation ($S_g = 1 - S_{org} - S_{gwirr}$), the SGOF table would look as seen in Table 9.15.

Again, it looks that we have no technical problem ending the table this way. But thinking the same way as the simulator, the maximum gas saturation in the system will be recognized as 0.433 instead of 0.740 in the original case. The effect of this modification on the initial volume in place and initial saturation distribution is shown in Table 9.16 and Fig. 9.22.

Table 9.14 The effect of Sw = 0.65 on evaluating the initial volumes in place and the initial saturation distribution when set to all points deeper than the given OWC

Fluid	Initial volume in place	Initial saturation range
Oil	1,606,274 stb	0.26–1.00
Water	658981stb	0.00–0.65
Free gas	190,050 Mscf	0.00–0.74

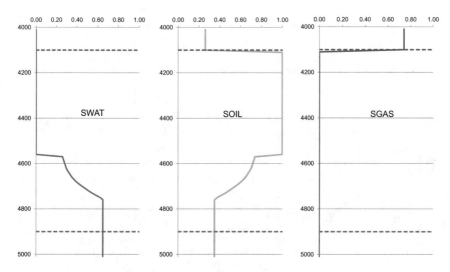

Fig. 9.21 Initial saturation profiles, violating SWOF end point rules

Table 9.15 SGOF

S_g	K_{rg}	K_{rog}	P_c
0.000	0.000000	0.841000	0.0000
0.020	0.000000	0.748674	0.0000
0.050	0.006145	0.623689	0.0000
0.100	0.024128	0.449028	0.0000
0.150	0.052081	0.310298	0.0000
0.200	0.094997	0.196275	0.0000
0.250	0.152222	0.114033	0.0000
0.300	0.228387	0.061589	0.0000
0.350	0.325759	0.029673	0.0000
0.400	0.467025	0.012494	0.0000
0.433	0.637678	0.000000	0.0000

Table 9.16 The effect of maximum gas saturation modification (0.433 instead of 0.740) on the initial volume in place and initial saturation distribution

Fluid	Initial volume in place	Initial Ssaturation range
Oil	1,175,107 stb	0.00–0.74
Water	1,241,147 stb	0.26–1.00
Free gas	113,003 Mscf	1.0–0.43

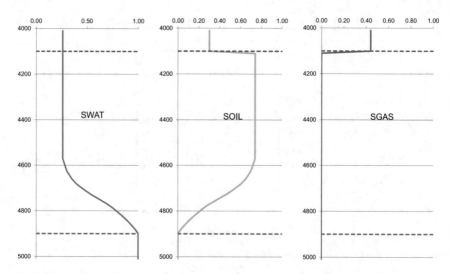

Fig. 9.22 Initial saturation profiles, violating SGOF end point rules

9.9 Preparing Saturation Functions (An Adopted Methodology)

The saturation functions for reservoir simulation work are prepared from the available SCAL laboratory experiments' results. Generally, there is no one standard methodology to prepare these functions. The requirements of the simulator and how it interprets and uses the given data, as described in the preceding sections, will dictate such methodology. It is to be mentioned that any methodology followed assumes the availability of reasonable representative SCAL data. However, this is not always the case. If SCAL data are not available for the studied reservoir, such data from analogue reservoirs could be used. The analogue reservoirs data can also be used to enrich limited data from studied reservoir. In the worst case of limited or no SCAL data and no analogue reservoir data, some correlation may be used. The analogue reservoirs should be selected considering similarity of depositional environments, porosity and permeability ranges and poro-perm general trends. Figure 9.23 is a decision tree for the work related to preparing saturation functions.

In this section, an adopted methodology to prepare the saturation functions for reservoir simulation work is described. The work flow of this adopted method is shown in Fig. 9.24. However, it is additionally recommended to go through the following detailed steps (supported by real field data example).

1. Starting with the available SCAL data (from studied reservoir, from analogue reservoirs or combination of both), perform quality checking to exclude erroneous and suspected data. The RE should check the SCAL lab reports for any issues regarding the laboratory experiments. On the other hand, he/she should carefully examine the reported data to exclude those experiments which

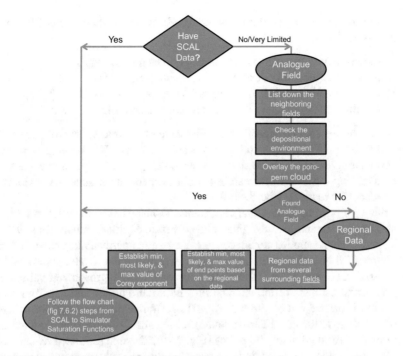

Fig. 9.23 Preparing saturation funstions for simulator decision tree

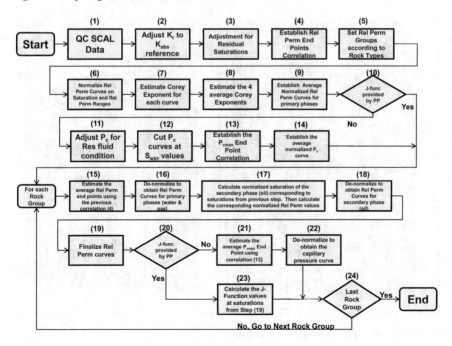

Fig. 9.24 SCAL to simulator saturation functions workflow

do not follow the physical understanding. Examples of those include (but not necessarily limited to):

– Experiment with relative permeability values greater than 1.
– Relative permeability curves bending down (Fig. 9.25).
– Convex relative permeability curve (Corey exponent less than 1).
– Capillary pressure curve that crosses other curves (Fig. 9.26).

2. Adjust the reported Relative Permeability values to be referenced to the absolute permeability (K_{abs}). Refer to Eqs. 9.1.2 and 9.1.3. The Klinkenberg corrected permeability should be used as the absolute permeability. In the case of Klinkenberg corrected permeability is not available and cannot be evaluated, permeability to air may be considered.

3. Adjust for residual saturations uncertainty. In many cases, the relative permeability experiments do not give reliable residual/critical values (Fig. 9.27). This especially applies for unsteady state relative permeability experiments. Figure 9.28 is a plot of the ratio K_{row}/K_{rw} (Logarithmic Scale) versus Water Saturation (Linear Scale). The plotted curve tends to approach vertical line to give a first estimate of the residual oil saturation. Further tuning of this estimation is carried out by plotting K_{row} (Logarithmic Scale) versus normalized oil saturation (Reversed Linear Scale). Using the first estimated residual oil saturation resulted in the red points (Fig. 9.29). The first estimated residual oil saturation is then modified until the best straight line is obtained (green points).

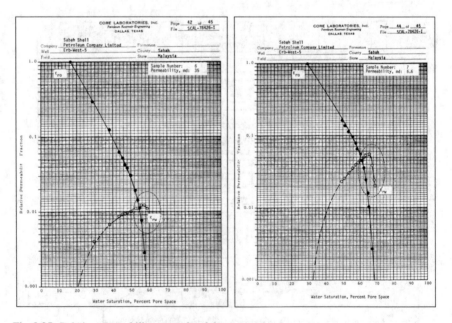

Fig. 9.25 Relative permeability curves bend down

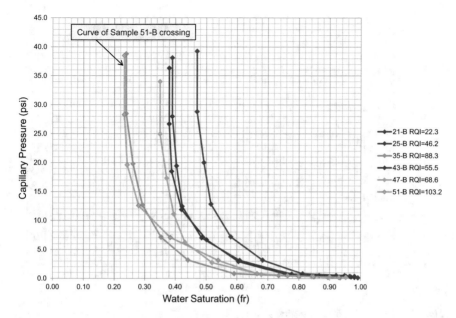

Fig. 9.26 Capillary pressure curves crossing

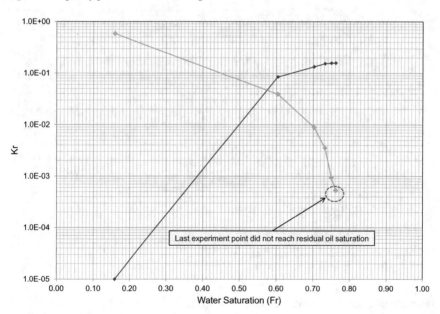

Fig. 9.27 Relative permeability water oil system

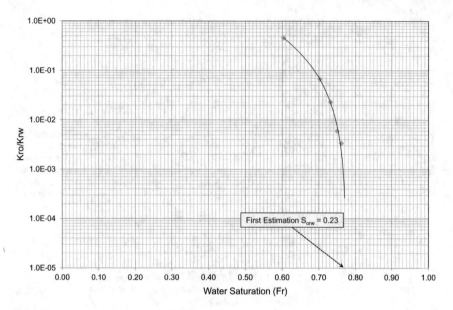

Fig. 9.28 Residual oil first estimation K_{row}/K_{rw} versus S_w

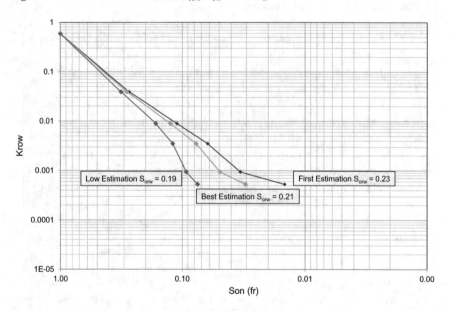

Fig. 9.29 Residual oil estimation tuning K_{row} versus S_{on}

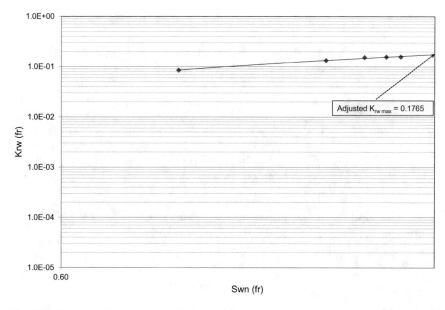

Fig. 9.30 Maximum K_{rw} adjustment K_{rw} versus S_{wn}

The residual oil saturation that gives such a straight line should be the best estimated one. The maximum K_{rw} is to be adjusted accordingly (Fig. 9.30).

Similar procedure is to be used to estimate more representative values for other critical/residual saturations. Figure 9.31 plots the ratio K_{rg}/K_{rog} (Logarithmic Scale) versus Gas Saturation (Linear Scale). Once again, the plotted curve tends to approach vertical line suggesting critical gas saturation (S_{gc}) of 0.01 (1.0%). Practically there is no need to more tune this small value.

4.   Establish Relative Permeability End Points Correlation. From the available laboratory experiments, the reported/estimated values of the above end points are to be plotted versus rock quality parameters. Porosity (Φ), permeability (K) and rock quality index (RQI) are the most common rock quality parameters that can be used for this purpose. In some cases, relying on availability, the pore size distribution can be used. It is advised that the reservoir engineer should try all possible parameters and chose the one which shows best correlation. Note that permeability (K) and rock quality index (RQI) are expected to show better presentation on logarithmic scale. However, both linear and logarithmic scales are to be tried. Also note that the residual and critical saturation values should decrease as the rock quality increases while the maximum relative permeability ($K_{r\,max}$) values should increase as the rock quality increases. Where ever the plotted data do not respect the expected trend, an average value could be adequate. For the given example (Figs. 9.32, 9.33, 9.34, 9.35, 9.36 and 9.37), the end points are plotted versus RQI.

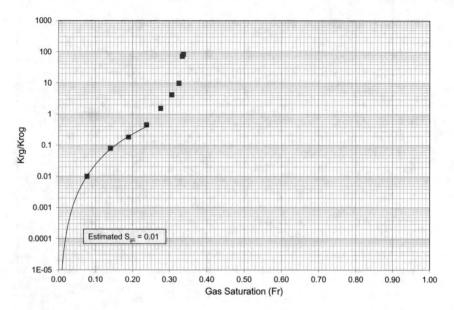

Fig. 9.31 Estimated critical gas saturation K_{rg}/K_{rog} versus S_g

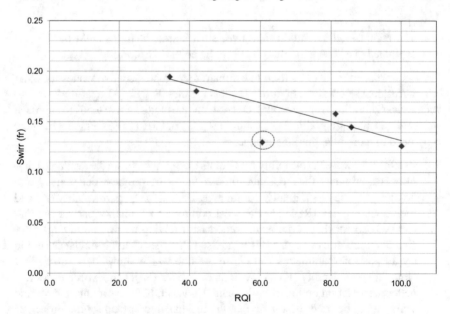

Fig. 9.32 End point correlation S_{wirr} versus RQI

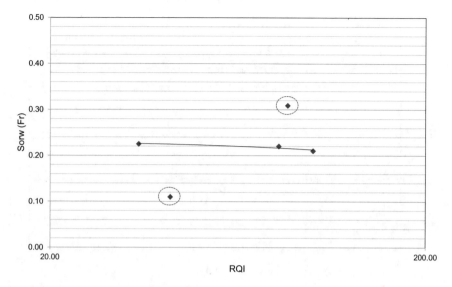

Fig. 9.33 End points correlation S_{orw} versus RQI

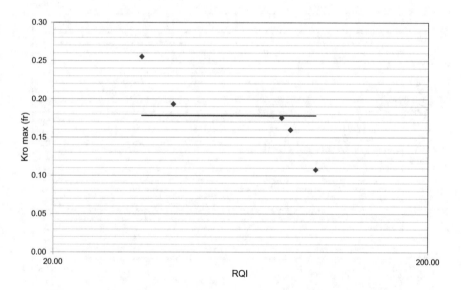

Fig. 9.34 End points correlation $K_{rw\ max}$

5. Set relative permeability groups according to the given rock types (from the
    static model work). It is preferred that each rock group has distinct range of
    the used end points correlating parameter (Figs. 9.38, 9.39, 9.40, 9.41, 9.42 and
    9.43). Otherwise the RE should consider subgrouping within each rock type.

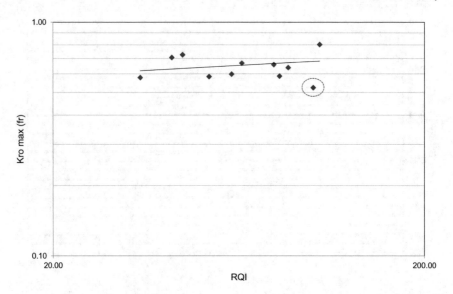

Fig. 9.35 End points correlation $K_{ro\,max}$ versus RQI

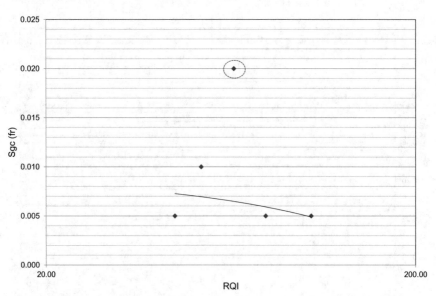

Fig. 9.36 End points correlation S_{gc} versus RQI

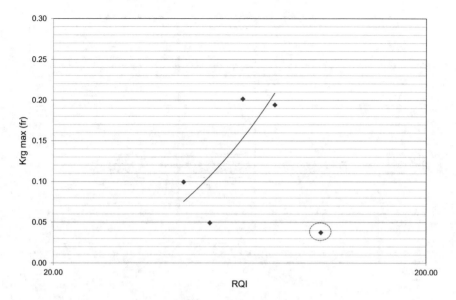

Fig. 9.37 End points correlation $K_{rg\,max}$ versus RQI

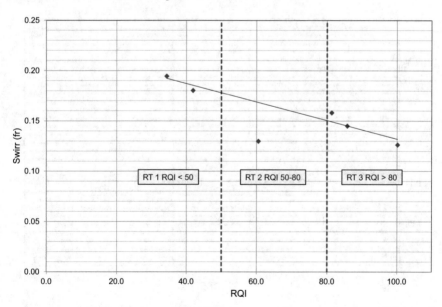

Fig. 9.38 End points correlation S_{wirr} versus RQI

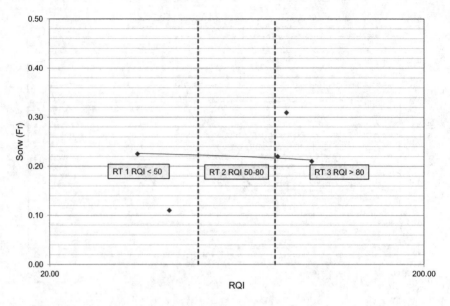

Fig. 9.39 End points correlation S_{orw} versus RQI

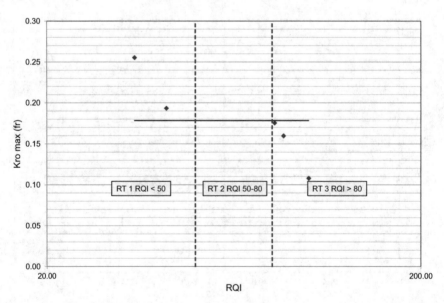

Fig. 9.40 End points correlation $K_{rw\,max}$ versus RQI

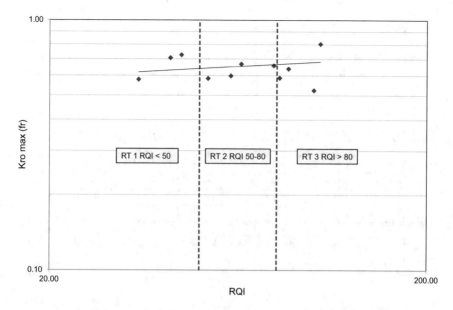

Fig. 9.41 End points correlation $K_{ro\,max}$ versus RQI

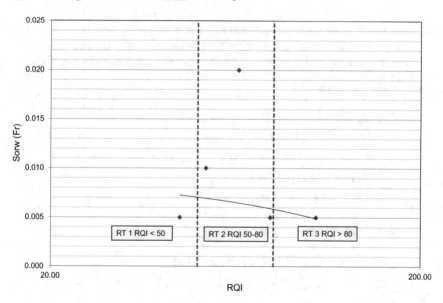

Fig. 9.42 End points correlation S_{gc} versus RQI

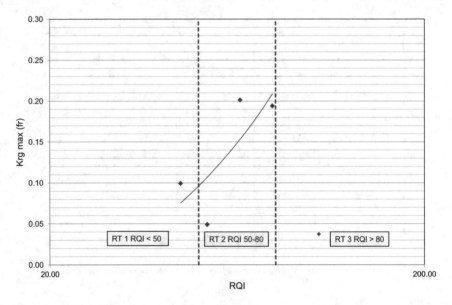

Fig. 9.43 End points correlation $K_{rg\,max}$ versus RQI

6. Normalize each relative permeability curve on both saturation and relative permeability ranges. It is highly recommended to normalize each phase saturation independently and according to its own saturation range as follows:

For water–oil system

$$S_w^* = (S_w - S_{wcr})/(1 - S_{orw} - S_{wcr})$$
$$S_o^* = (S_o - S_{orw})/(1 - S_{orw} - S_{wirr})$$

(Note that in the case of $S_{wirr} = S_{wcr}$, then $S_o^* = 1 - S_w^*$)

$$Krw^* = Krw/(Krw\,max)$$
$$Krow^* = Krow/(Krow\,max)$$

For gas-oil system

$$S_g^* = (S_g - S_{gc})/(1 - S_{org} - S_{wirr} - S_{gc})$$
$$S_o^* = (S_o - S_{org})/(1 - S_{org} - S_{wirr}) \text{ where, } S_o = 1 - S_g - S_{wirr}$$

(Note that in the case of $S_{gc} = 0$, then $S_o^* = 1 - S_g^*$)

$$K_{rg}^* = K_{rg}/(K_{rg\,max})$$
$$K_{rog}^* = K_{rog}/(K_{rog\,max})$$

Figures 9.44 and 9.45 show normalized relative permeability plots for water-oil system and gas-oil system respectively. Note that, for our case gas-oil system, only the relative permeability data to the gas phase are available.

7. Estimate Corey Exponent for each normalized curve.
8. Estimate the average 4 Corey Exponents.

    Remember that four different average exponents are to be estimated; N_o and N_w for water-oil system and N_g and N_o for gas-oil system. Note that N_o for water-oil system is not necessarily equal to N_o for gas-oil system. However, N_o for gas-oil system was considered as equal to N_o for water-oil system in the given example (Table 9.17) due to the absence of relative permeability data to oil for the gas-oil system. In the averaging procedure, it is not recommended to use any weighting factor because the confidence degree is expected to be the same for all curves (all samples). Using absolute permeability as a weighting factor is not a good practice. There is no reason to give more weight to the higher permeable samples. Table 9.17 shows the estimated Corey exponents for each sample and the average Corey exponent for each phase.

9. Establish average <u>normalized</u> relative permeability curve for the primary phase of each fluid system; water phase for the water–oil system and gas phase for the gas-oil system. As it is well known the formula to be used is:

$$K_{ri}^* = \left(S_i^*\right)^{Ni}, \text{ where i is the phase (w or g)}$$

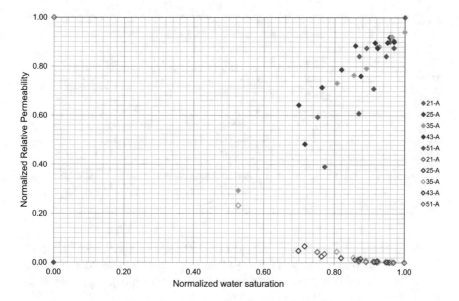

Fig. 9.44 Normalized relative permeability (Water–Oil)

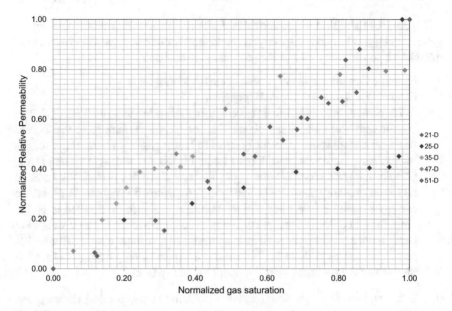

Fig. 9.45 Normalized relative permeability (Gas-Oil)

Table 9.17 Estimated corey exponents

Sample	Φ	K (md)	RQI	Water–Oil		Gas-Oil	
				N_w	N_o	N_g	N_o
21-A	0.261	458	41.9	1.9	2.6	–	–
25-A	0.357	304	34.4	1.2	2.4	–	–
35-A	0.319	2347	85.8	1.8	2.0	–	–
43-A	0.316	2088	81.3	2.3	2.1	–	–
51-A	0.291	2919	100.2	(3.7)	2.1	–	–
21-D	0.281	563	44.8	–	–	1.5	–
25-D	0.269	746	52.7	–	–	2.0	–
35-D	0.261	1083	64.4	–	–	(0.8)	–
47-D	0.302	1859	78.5	–	–	1.2	–
51-D	0.311	3374	104.2	–	–	1.3	–
Average Corey Exponent				1.8	2.2	1.5	2.2

Use reasonable S_i^* steps (e.g. 0.05) and cover the whole range of S_i^* (0.0 to 1.0) and naturally the whole range of K_{ri}^* (0.0 to 1.0). This step should end up with two normalized relative permeability curves for each rock group; K_{rw}^* versus S_w^* (Fig. 9.46) and K_{rg}^* versus S_g^* (Fig. 9.47).

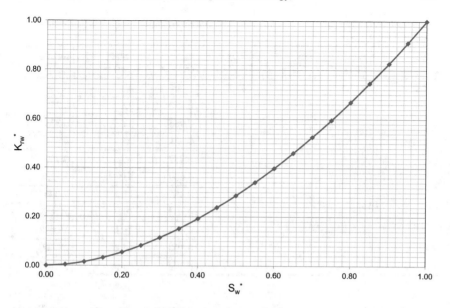

Fig. 9.46 Average normalized relative permeability (Water)

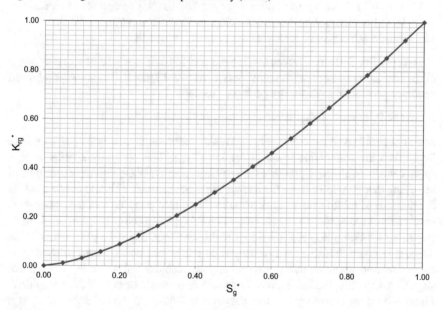

Fig. 9.47 Average normalized relative permeability (Gas)

Table 9.18 Adjusting capillary pressure to reservoir fluid condition oil brine experiment (Res P_c = 0.625 * lab P_c)

Sample	K (md)	Φ (fr)	RQI	S_w (fr)	Lab P_c (psi)	Res P_c (psi)
25-B	587.0	0.275	46.2	0.990	0.3	0.188
				0.980	0.4	0.250
				0.965	0.6	0.375
				0.921	0.8	0.500
				0.777	1.2	0.750
				0.609	5.0	3.125
				0.485	11.2	7.000
				0.422	19.9	12.438
				0.404	31.1	19.438
				0.390	44.8	28.000
				0.390	61.0	38.125

10.   If the J-function information is provided by the petrophysicist, proceed directly to step 15. If such information is not provided or it is preferred to prepare own P_c /J-Function saturation relationship, go to the next step.

11.   Adjust lab capillary pressure data to reservoir condition.

Refer to Eq. 8.4.1 and Tables 8.4.1, 8.4.2 and 8.4.3. Table 9.18 gives an example of the capillary pressure adjustment for one sample. Figure 9.48 shows the adjusted capillary pressure curves for the example problem.

12.   Cut P_c curves at S_{wirr} values.

In many capillary pressure experiments, the water saturation is forced to values lower than the irreducible water saturation (S_{wirr}). This specially applies to high speed centrifuge and high pressure mercury injection experiments. Any experimental recording that goes lower than the irreducible water saturation should not be used for the purpose of initial saturation distribution. It is also possible that the experiment reaches the irreducible water saturation before applying the maximum experiment pressure. Our example problem belongs to this category. Examining the data in Table 4.6.3, it is noted that the last two saturation entries are equal which suggests that the irreducible water saturation (0.390) already reached before the last experiment step. Consequently, the last capillary pressure step should be cut off from the curve. Figure 9.49 shows capillary pressure curves after cutting off the last experiment step.

13.   Establish the $P_{c\,max}$ end point correlation.

For this step we follow the same procedure used for establishing relative permeability end points (Step 4). Refer also to Fig. 9.50.

14.   Establish average normalized capillary pressure curve.

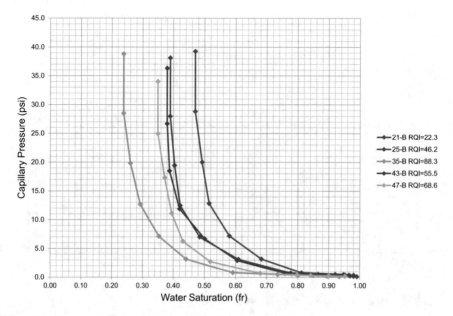

Fig. 9.48 Capillary pressure curves (adjusted to reservoir condition)

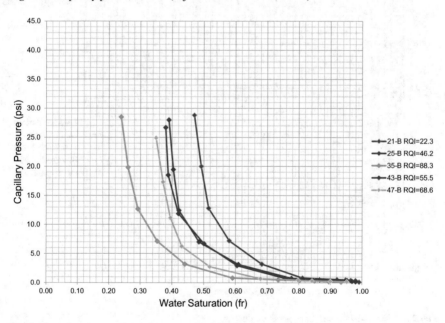

Fig. 9.49 Capillary pressure curves (cut @ irreducible water saturation)

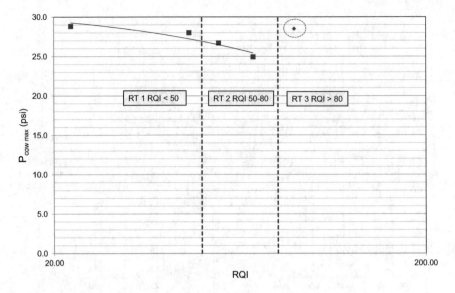

Fig. 9.50 End point correlation ($P_{\text{cow max}}$ versus RQI)

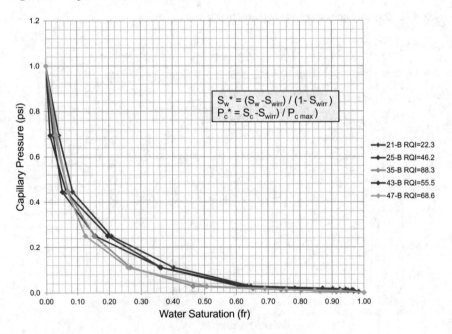

Fig. 9.51 Normalizing capillary pressure curves

Normalization is to be carried out on each available P_c curve (Fig. 9.51) and then the normalized curves are to be averaged in one curve (Fig. 9.52). When normalizing P_c curves, it should be noticed that the saturation range is not the same like the case of K_r curve normalization.

$$S_w^* = (S_w - S_{wirr})/(1 - S_{wirr})$$
$$P_c^* = P_c/P_{c\ max}$$

15. <u>For each rock group</u>, estimate the average relative permeability end points correlating parameter (porosity, permeability or rock quality index) using the established correlation from step 4. Consequently, estimate the relative permeability average end points. Refer back to Figs. 9.38, 9.39, 9.40, 9.41, 9.42 and 9.43. Table 9.19 shows the estimated end points.

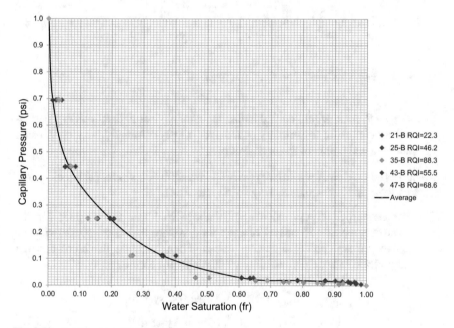

Fig. 9.52 Average normalized capillary pressure curve

Table 9.19 Relative permeability end points for different rock types

Rock Type	RQI	End points (fraction)						
		S_{wirr}	S_{orw}	S_{gcr}	S_{org}	$K_{ro\ max}$	$K_{rw\ max}$	$K_{rg\ max}$
1	<50	0.190	0.222	0.0070	0.222	0.62	0.1783	0.09
2	50–80	0.165	0.219	0.0072	0.219	0.65	0.1783	0.16
3	>80	0.135	0.213	0.0053	0.213	0.68	0.1783	0.22

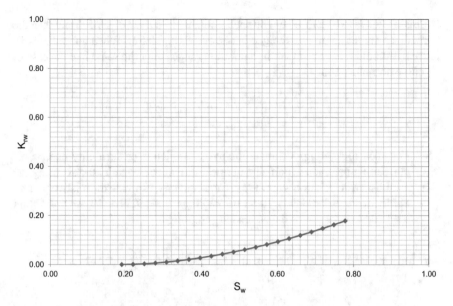

Fig. 9.53 De-normalized relative permeability (Water) RT 1

16. *For each rock group*, de-normalize the average normalized relative perme-
     ability curves for the primary phases (previously obtained from step 9) using
     suitable end points of the rock group.

     This step should end up with two relative permeability curves for each rock group;
     K_{rw} versus S_w and K_{rg} versus S_g (Figs. 9.53 and 9.54).

17. For each rock group, calculate the normalized saturation values for the
     secondary phase (oil phase) of each fluid system that corresponds with the
     above primary phase saturation values. Then calculate the corresponding
     normalized relative permeability values using the formulae:

$$K_{row}^* = \left(S_o^*\right)^{No}, \text{ (water-oil system)}$$
$$K_{rog} = \left(S_o^*\right)^{No}, \text{ (gas-oil system)}$$

     Tables 9.20 and 9.21 explain the calculation procedure for oil-water system and
     gas-oil system respectively. Figure 9.55 shows the water-oil system normalized rela-
     tive permeability curves. Note that the normalized relative permeability curve for the
     water phase (blue curve) is the average normalized curve obtained in step 9. Similarly,
     Fig. 9.56 shows the gas-oil system normalized relative permeability curves.

18. For each rock group, de-normalize the values calculated in the previous step
     using suitable end points of the rock group. This step should end up with two

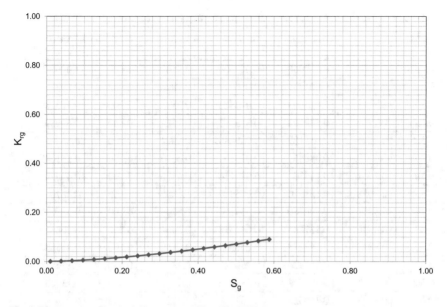

Fig. 9.54 De-normalized relative permeability (Gas) RT 1

relative permeability curves for each rock group; K_{row} versus S_w (water–oil system) and K_{rog} versus S_g (gas-oil system). Note that it is not needed to de-normalize oil saturation for this step.

The outcomes from this step and the outcome from step 16 represent the water-oil system relative permeability (Table 9.22 and Fig. 9.57) and the gas-oil relative permeability (Table 9.23 and Fig. 9.58).

19. For each rock group, finalize the obtained relative permeability curves to satisfy the simulator requirements (refer back to Sect. 9.8). To fulfill this goal, it is needed to add more entries to the obtained relative permeability tables. The entries shown in red in the previous two tables (Tables 9.22 and 9.23) have been added for this purpose. The dashed parts of the relative permeability curves (Figs. 9.57 and 9.58) are the graphical presentation of those added entries.

20. If the J-function information is provided by the petrophysicist, proceed to step 23.

21. *For each rock group*, use the correlation from step 13 ($P_{c\,max}$ end point correlation) to estimate the average $P_{c\,max}$ (Table 9.24).

22. For each rock group, de-normalize the average normalized P_c curve using proper end points ($P_{c\,max}$ and S_{wirr}) to obtain the capillary pressure curve for the rock group. Refer to Table 9.25 and Fig. 9.59.

Go directly to step 24.

Table 9.20 Normalized relative permeability to oil (Water–Oil System)

S_w^*	S_w	K_{rw}^*	K_{rw}	S_o	S_o^*	K_{row}^* ($N_o = 2.2$)
0.00	0.1900	0.000000	0.000000	0.8100	1.000	1.000000
0.05	0.2194	0.004551	0.000812	0.7806	0.950	0.893289
0.10	0.2488	0.015849	0.002826	0.7512	0.900	0.793110
0.15	0.2782	0.032882	0.005863	0.7218	0.850	0.699394
0.20	0.3076	0.055189	0.009840	0.6924	0.800	0.612066
0.25	0.3370	0.082469	0.014704	0.6630	0.750	0.531049
0.30	0.3664	0.114503	0.020414	0.6336	0.700	0.456263
0.35	0.3958	0.151120	0.026945	0.6042	0.650	0.387623
0.40	0.4252	0.192180	0.034266	0.5748	0.600	0.325037
0.45	0.4546	0.237565	0.042358	0.5454	0.550	0.268410
0.50	0.4840	0.287175	0.051203	0.5160	0.500	0.217638
0.55	0.5134	0.340920	0.060786	0.4866	0.450	0.172611
0.60	0.5428	0.398724	0.071092	0.4572	0.400	0.133209
0.65	0.5722	0.460515	0.082110	0.4278	0.350	0.099300
0.70	0.6016	0.526231	0.093827	0.3984	0.300	0.070740
0.75	0.6310	0.595813	0.106234	0.3690	0.250	0.047366
0.80	0.6604	0.669209	0.119320	0.3396	0.200	0.028991
0.85	0.6898	0.746370	0.133078	0.3102	0.150	0.015396
0.90	0.7192	0.827250	0.147499	0.2808	0.100	0.006310
0.95	0.7486	0.911806	0.162575	0.2514	0.050	0.001373
1.00	0.7780	1.00000	0.178300	0.2220	0.000	0.000000

23. For each rock group, the J-function values to be calculated for different water saturations using the provided formula. Table 9.26 shows an example of calculating the J-function using a formula provided by petrophysicist. Note the last J-function entry has been set to zero to honor the zero entry pressure reported for all capillary pressure experiments.

For some cases, the petrophysicist provided formula may relates the J-function to normalized water saturation (S_w^*) rather than the water saturation (S_w). Be careful, that the normalized water saturation used for J-function formulation is not the same as the one used in the water-oil relative permeability treatment (Refer to step 14).

24. Go back to step 15 and repeat till covering all rock groups.

In the above procedure, only the water–oil system drainage capillary pressure was considered. The imbibition water–oil capillary pressure can be treated the same way considering the different range of saturation (Fig. 9.12). However, the availability of the imbibition capillary pressure is very rare. To fulfill the requirements of the reservoir simulation model, it is suggested to construct an approximate imbibition

Table 9.21 Normalized relative permeability to oil (Gas-Oil System)

S_g^*	S_g	K_{rg}^*	K_{rg}	S_o	S_o^*	K_{rog}^* $(N_o = 2.2)$
0.00	0.0100	0.000000	0.000000	0.8000	0.9830	0.9630
0.05	0.0389	0.011180	0.001006	0.7711	0.9338	0.8602
0.10	0.0678	0.031623	0.002846	0.7422	0.8847	0.7637
0.15	0.0967	0.058095	0.005229	0.7133	0.8355	0.6735
0.20	0.1256	0.089443	0.008050	0.6844	0.7864	0.5894
0.25	0.1545	0.125000	0.011250	0.6555	0.7372	0.5114
0.30	0.1834	0.164317	0.014789	0.6266	0.6881	0.4394
0.35	0.2123	0.207063	0.018636	0.5977	0.6389	0.3733
0.40	0.2412	0.252982	0.022768	0.5688	0.5898	0.3130
0.45	0.2701	0.301869	0.027168	0.5399	0.5406	0.2585
0.50	0.2990	0.353553	0.031820	0.5110	0.4915	0.2096
0.55	0.3279	0.407891	0.036710	0.4821	0.4423	0.1662
0.60	0.3568	0.464758	0.041828	0.4532	0.3932	0.1283
0.65	0.3857	0.524047	0.047164	0.4243	0.3440	0.0956
0.70	0.4146	0.585662	0.052710	0.3954	0.2949	0.0681
0.75	0.4435	0.649519	0.058457	0.3665	0.2457	0.0456
0.80	0.4724	0.715542	0.064399	0.3376	0.1966	0.0279
0.85	0.5013	0.783661	0.070530	0.3087	0.1474	0.0148
0.90	0.5302	0.853815	0.076843	0.2798	0.0983	0.0061
0.95	0.5591	0.925945	0.083335	0.2509	0.0491	0.0013
1.00	0.5880	1.000000	0.090000	0.2220	0.0000	0.0000

capillary pressure curve. Such synthesized curve is constructed by pulling the available drainage capillary pressure curve tail from ($S_w = 1.0$) to ($S_w = 1.0 - S_{orw}$). Refer to Fig. 9.60. This is can be performed as follows;

- Normalize the available drainage capillary pressure curve (water saturation only) along the saturation span (from $S_w = S_{wirr}$ to $S_w = 1.0$).
- De-normalize the obtained normalized water saturation considering the span (from $S_w = S_{wirr}$ to $S_w = 1.0 - S_{orw}$).
- Resample the P_c values at the original S_w values using simple averaging.

Refer to Tables 9.27 and 9.28 and Fig. 9.61 for clarification of this procedure.

Gas-oil capillary pressure can be handled in a similar way as the water–oil capillary pressure. It is understood that the gas-oil capillary pressure values are very small as compared to the water–oil capillary pressure values (refer to Tables 9.3 and 9.4). Consequently, it is an acceptable practice to simply set all gas-oil capillary pressure entries in the SGOF tables to zero.

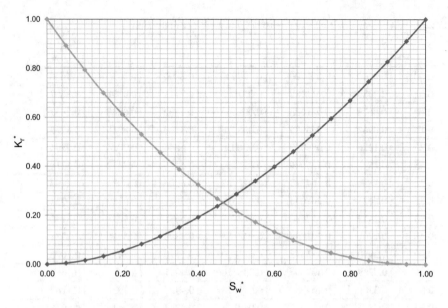

Fig. 9.55 Normalized relative permeability (Water–Oil), RT 1

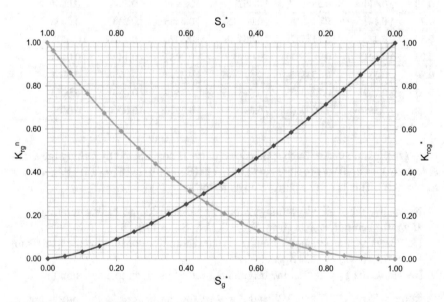

Fig. 9.56 Normalized relative permeability (Gas-Oil), RT 1

Table 9.22 De-normalized relative permeability (Water–Oil System)

S_w^*	S_w	K_{rw}^*	K_{rw}	S_o	S_o^*	K_{row}^*	K_{row}
0.00	0.1900	0.000000	0.000000	0.8100	1.000	1.000000	0.620000
0.05	0.2194	0.004551	0.000812	0.7806	0.950	0.893289	0.553839
0.10	0.2488	0.015849	0.002826	0.7512	0.900	0.793110	0.491728
0.15	0.2782	0.032882	0.005863	0.7218	0.850	0.699394	0.433624
0.20	0.3076	0.055189	0.009840	0.6924	0.800	0.612066	0.379481
0.25	0.3370	0.082469	0.014704	0.6630	0.750	0.531049	0.329251
0.30	0.3664	0.114503	0.020414	0.6336	0.700	0.456263	0.282883
0.35	0.3958	0.151120	0.026945	0.6042	0.650	0.387623	0.240326
0.40	0.4252	0.192180	0.034266	0.5748	0.600	0.325037	0.201523
0.45	0.4546	0.237565	0.042358	0.5454	0.550	0.268410	0.166414
0.50	0.4840	0.287175	0.051203	0.5160	0.500	0.217638	0.134935
0.55	0.5134	0.340920	0.060786	0.4866	0.450	0.172611	0.172611
0.60	0.5428	0.398724	0.071092	0.4572	0.400	0.133209	0.133209
0.65	0.5722	0.460515	0.082110	0.4278	0.350	0.099300	0.099300
0.70	0.6016	0.526231	0.093827	0.3984	0.300	0.070740	0.070740
0.75	0.6310	0.595813	0.106234	0.3690	0.250	0.047366	0.047366
0.80	0.6604	0.669209	0.119320	0.3396	0.200	0.028991	0.028991
0.85	0.6898	0.746370	0.133078	0.3102	0.150	0.015396	0.015396
0.90	0.7192	0.827250	0.147499	0.2808	0.100	0.006310	0.006310
0.95	0.7486	0.911806	0.162575	0.2514	0.050	0.001373	0.001373
1.00	0.7780	1.00000	0.178300	0.2220	0.000	0.000000	0.000000
	0.80		0.20				0.00
	0.85		0.26				0.00
	0.90		0.36				0.00
	0.95		0.60				0.00
	1.00		1.00				0.00

9.10 Hysteresis

The term **"hysteresis"** refers to irreversibility or path dependence. It was coined around by Sir James Alfred Ewing (1889) to describe the behavior of magnetic materials. For multiphase flow in porous media, hysteresis manifests itself through the dependence of relative permeability and capillary pressure not only on the saturation value but also on the saturation path. Accordingly, relative permeability and capillary pressures are not unique functions of saturation. Let us consider relative permeability to elaborate on this phenomenon. The relative permeability to one phase has different values at the same phase saturation depending on the phase saturation change direction. At the same phase saturation, the relative permeability to the phase is higher

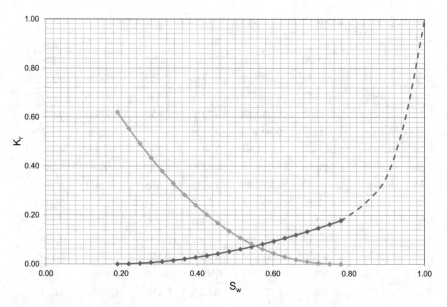

Fig. 9.57 De-normalized relative permeability (Water–Oil) RT 1

if its saturation increasing rather than if it is decreasing. Subsequently, two sets of relative permeability curves are needed. These are drainage curves (non-wetting phase saturation increasing) and imbibition curves (non-wetting phase saturation decreasing).

As previously mentioned in Sect. 9.6, the flow process will shuttle between the two curves according to the direction of saturation change. Moving from one curve to another the flow process follows scanning curves.

Figure 9.62, shows typical drainage and imbibition relative permeability curves for the non-wetting phase. The drainage curve describes the process of non-wetting phase saturation increasing. This process follows the path (1) to (2) from drainage critical/residual non-wetting phase saturation (S_{ncrd}/S_{nrd}) to maximum non-wetting phase saturation (S_{nmax}). The imbibition curve describes the process of non-wetting phase saturation decreasing. This process follows the path (2) to (3) from the maximum non-wetting phase saturation to imbibition critical/residual non-wetting phase saturation (S_{ncri}/S_{nri}). Suppose that the drainage process (non-wetting phase saturation increasing) is reversed at some intermediate point (4). The process will not follow the same curve in the opposite direction. Rather, it will follow a scanning curve (4) to (5) from the intermediate non-wetting phase saturation to trapped critical non-wetting saturation (S_{ntcr}). The same analysis can be performed for the wetting phase (Fig. 9.63). There are different options built in Eclipse simulator to construct the scanning curves. These are Carlson's method, Killough's method and Jargon's method.

Table 9.23 De-normalized relative permeability (Gas-Oil System)

S_g^*	S_g	K_{rg}^*	K_{rg}	S_o	S_o^*	K_{rog}^*	K_{rog}
	0.00		0.00				0.62
0.00	0.0100	0.000000	0.000000	0.8000	0.9830	0.9630	0.597039
0.05	0.0389	0.011180	0.001006	0.7711	0.9338	0.8602	0.533328
0.10	0.0678	0.031623	0.002846	0.7422	0.8847	0.7637	0.473518
0.15	0.0967	0.058095	0.005229	0.7133	0.8355	0.6735	0.417565
0.20	0.1256	0.089443	0.008050	0.6844	0.7864	0.5894	0.365427
0.25	0.1545	0.125000	0.011250	0.6555	0.7372	0.5114	0.317057
0.30	0.1834	0.164317	0.014789	0.6266	0.6881	0.4394	0.272407
0.35	0.2123	0.207063	0.018636	0.5977	0.6389	0.3733	0.231426
0.40	0.2412	0.252982	0.022768	0.5688	5898	0.3130	0.194060
0.45	0.2701	0.301869	0.027168	0.5399	05,406	0.2585	0.160251
0.50	0.2990	0.353553	0.031820	0.5110	0.4915	0.2096	0.129938
0.55	0.3279	0.407891	0.036710	0.4821	0.4423	0.1662	0.103055
0.60	0.3568	0.464758	0.041828	0.4532	0.3932	0.1283	0.079531
0.65	0.3857	0.524047	0.047164	0.4243	0.3440	0.0956	0.059286
0.70	0.4146	0.585662	0.052710	0.3954	0.2949	0.0681	0.042235
0.75	0.4435	0.649519	0.058457	0.3665	0.2457	0.0456	0.028279
0.80	0.4724	0.715542	0.064399	0.3376	0.1966	0.0279	0.017309
0.85	0.5013	0.783661	0.070530	0.3087	0.1474	0.0148	0.009192
0.90	0.5302	0.853815	0.076843	0.2798	0.0983	0.0061	0.003767
0.95	0.5591	0.925945	0.083335	0.2509	0.0491	0.0013	0.000820
1.00	0.5880	1.000000	0.090000	0.2220	0.0000	0.0000	0.000000
	0.60		0.095				0.00
	0.65		0.140				0.00
	0.70		0.260				0.00
	0.75		0.500				0.00
	0.81		0.800				0.00

Generally speaking, application of hysteresis is not needed to study normal reservoir depletion processes. However, it should be considered where reverse saturation change is expected due to some reservoir operations. These operations include water flooding of gas invaded zone, migration of oil into gas cap (gas cap desaturation) and collapse and return of water cone (cyclic production). It is of special importance in studying water alternating gas (WAG) operations.

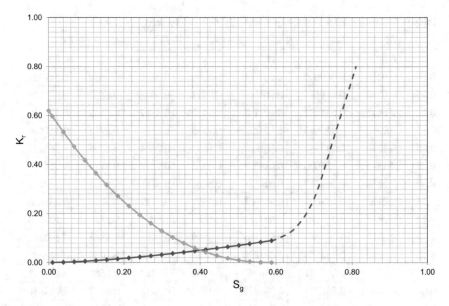

Fig. 9.58 De-normalized relative permeability (Gas-Oil) RT 1

Table 9.24 $P_{c\,max}$ end point for different rock types

Rock type	RQI	$P_{c\,max}$ (psi)
1	<50	28.5
2	50–80	25.8
3	>80	23.0

9.11   Electrical Rock Properties

Electrical reservoir rock properties are applicable in open-hole log analysis aiming at estimating water saturation in the reservoir rock. As so, the main interest to these properties is by petrophysicist. Being highly interested in water saturation estimation, the reservoir engineer should have reasonable understanding about these properties. This section will give simple explanation about the basics of the electric reservoir rock properties and their related application to estimate water saturation.

Conductivity (C) of any material is the measure of the ability of this material to transfer electric current. The unit used to measure material conductivity is mho/meter or m-mho/meter (milli-mho/meter). The more popular electrical property is the resistivity. **Resistivity (R)** of any material is the measure of the ability of this material not to allow transfer of electric current or in other words the ability to resist the transfer of electric current. Resistivity is the reciprocal of conductivity and it is measured in units of ohm. Meter.

Any reservoir rock consists of solid grains (matrix) packed in some way allowing void space which is filled with reservoir fluids. The matrix is generally consisting of

Table 9.25 De-normalized capillary pressure (Water–Oil System)

S_w	$S_w{}^*$	$P_{cow}{}^*$	P_{cow} (psi)
0.1900	0.0000	1.0000	28.500
0.2194	0.0363	0.5440	15.504
0.2488	0.0726	0.4275	12.184
0.2782	0.1089	0.3624	10.328
0.3076	0.1452	0.3120	8.892
0.3370	0.1815	0.2678	7.632
0.3664	0.2178	0.2252	6.418
0.3958	0.2541	0.1868	5.324
0.4252	0.2904	0.1580	4.503
0.4546	0.3267	0.1311	3.736
0.4840	0.3630	0.1085	3.092
0.5134	0.3993	0.0905	2.597
0.5428	0.4356	0.0770	2.195
0.5722	0.4719	0.0650	1.853
0.6016	0.5081	0.0539	1.536
0.6310	0.5444	0.0446	1.271
0.6604	0.5807	0.0349	0.995
0.6898	0.6170	0.0269	0.767
0.7192	0.6533	0.0209	0.596
0.7486	0.6896	0.0202	0.576
0.7780	0.7259	0.0197	0.561
0.8000	0.7531	0.0195	0.556
0.8500	0.8148	0.0184	0.524
0.9000	0.8765	0.0159	0.453
0.9500	0.9383	0.0112	0.319
1.0000	1.0000	0.0000	0.000

minerals like quartz (sandstone), calcite (carbonate) and different clay minerals. The void space is filled with hydrocarbons (oil and/or gas) and brine (formation water). Except for some clay minerals, the rock matrix does not allow electric current transmission or in other words it is non-conductive. The same is applicable to hydrocarbons. The brine (formation water) allows transmission of electric current. Its conductivity relies on the amount of solids dissolved in it or in other words its salinity. From this clarification, it can be concluded that the rock ability to transmit electric current is a function of the amount of formation water in the pore space (in other words porosity and formation water saturation), formation water salinity and rock tortuosity. This understanding represents the basis used by scholars to develop methods for estimating water saturation from the electric reservoir rock properties.

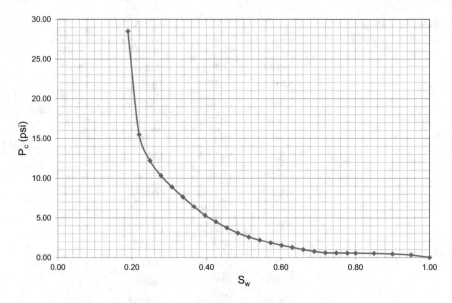

Fig. 9.59 De-normalized capillary pressure (Water–Oil), RT 1

Let us now set some definitions to help in this analysis. **Formation Resistivity Factor (F)** is defined as the ratio between the resistivity of fully water saturated rock to the resistivity of the water occupying the pore space:

$$F = R_o/R_w \tag{9.11.1}$$

where,

$R_w =$ Resistivity of the formation water.
$R_o =$ Resistivity of the 100% water saturated rock.

It is obvious that the factor F is always greater than 1.0

Archie (1942) formula describes the relation between the formation resistivity factor (F) and formation porosity (Φ). For clean formation, the formula is:

$$F = a/\Phi^m \tag{9.11.2}$$

where,

$a =$ Archie Constant or Tortuosity Factor.
$m =$ Cementation Factor.

The two constants a and m in formula 9.11.2 can be estimated in laboratory as a part of SCAL program. Standard value used for a is 1.0 while standard value used for m is around 2.0 Different useful forms of Archie formula are Humble formula for sands:

Table 9.26 J-function (Water–Oil System)

$J = 0.1 \, S_w^{-3.8}$

S_w	J-function
0.1900	55.047
0.2194	19.759
0.2488	12.925
0.2782	8.824
0.3076	6.237
0.3370	4.539
0.3664	3.385
0.3958	3.385
0.4252	2.578
0.4546	2.000
0.4840	1.576
0.5134	1.260
0.5428	1.019
0.5722	0.834
0.6016	0.690
0.6310	0.575
0.6604	0.484
0.6898	0.410
0.7192	0.350
0.7486	0.301
0.7780	0.260
0.8000	0.233
0.8500	0.185
0.9000	0.149
0.9500	0.122
1.0000	0.000

$$F = 0.62/\Phi^{2.15} \tag{9.11.3}$$

Or in simpler form

$$F = 0.81/\Phi^{2.0} \tag{9.11.4}$$

and Archie formula for compact formation:

$$F = 1.0/\Phi^{2} \tag{9.11.5}$$

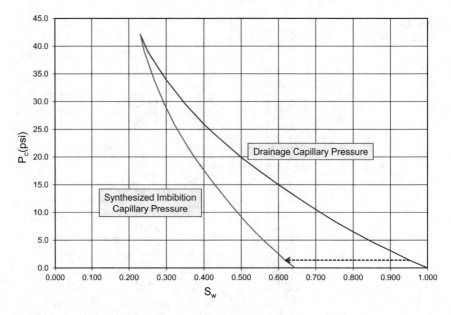

Fig. 9.60 Synthesized imbibition capillary pressure curve

To estimate the constants a and m for a reservoir formation, let us consider Archie formula (9.11.2) and we can write,

$$\text{Log } F = \text{Log } a - m \text{ Log } \Phi$$

This is a linear relation on a Log–Log plot. To obtain the factors a and m, values of the formation resistivity factor (F, formula 9.11.1) and porosity (Φ) from laboratory experiments are to be plotted on Log–Log scale. The slope of the fitted straight line (Fig. 9.64) gives the value of m while the intercept of the line at porosity of 1.0 gives the value of a.

Resistivity Index (RI) is defined as the ratio between the true formation rock (saturated with both hydrocarbon and formation water) resistivity and resistivity of the same rock if 100% saturated with formation water only.

$$RI = R_t/R_o \tag{9.11.6}$$

where,

$R_t =$ True resistivity of the rock system at some water saturation value.
$R_o =$ Resistivity of the 100% water saturated rock.

According to the above explanation, the resistivity index (RI) is always greater than 1.0.

Table 9.27 Synthesized imbibition capillary pressure curve

	Drainage			Synthesized imbibition	
	S_w	P_c	$S_w{}^*$	S_w	P_c
S_{wirr}	0.230	42.2	0.000	0.230	42.2
	0.250	39.1	0.026	0.241	39.1
	0.300	33.9	0.091	0.268	33.9 33.9
	0.350	29.6	0.156	0.295	29.6
	0.400	25.9	0.221	0.321	25.9
	0.450	22.8	0.286	0.348	22.8
	0.500	19.9	0.351	0.375	19.9
	0.550	17.4	0.416	0.402	17.4
	0.600	15.0	0.481	0.429	15.0
$1-S_{orw}$	0.644	13.0	0.538	0.453	13.0
	0.700	10.5	0.610	0.483	10.5
	0.750 0.750	8.4	0.675	0.510	8.4 8.4
	0.800	6.5	0.740	0.536	6.5
	0.850	4.7	0.805 0.805	0.563	4.7
	0.900	3.1	0.870 0.870	0.590	3.1
	0.950	1.4	0.935	0.617	1.4
	1.000	0.0	1.000	0.644	0.0

The useful relation between the water saturation (S_w) and resistivity index for clean formation was also suggested by Archie (1942) as;

$$RI = 1.0/(S_w)^n \qquad (9.11.7)$$

or

$$S_w = (R_o/R_t)^{1/n} \qquad (9.11.8)$$

where,

n =   Saturation Exponent.

To estimate the value of the constant n, a procedure similar to the one used for estimating a and m is to be applied. Considering the formula (9.11.7), we can write;

$$Log\ RI = -n\ Log(S_w)$$

	Synthesized imbibition capillary pressure		Resampled synthesized capillary pressure	
Table 9.28 Synthesized imbibition capillary pressure	S_w	P_c	S_w	P_c
	0.230	42.2	0.230	42.2
	0.241	39.1		
			0.250	37.3
	0.268	33.9		
	0.295	29.6		
			0.300	28.8
	0.321	25.9		
	0.348	22.8		
			0.350	22.6
	0.375	19.9		
			0.400	17.6
	0.402	17.4		
	0.429	15.0		
			0.450	13.2
	0.453	13.0		
	0.483	10.5		
			0.500	9.1
	0.510	8.4		
	0.536	6.5		
			0.550	5.6
	0.563	4.7		
	0.590	3.1		
			0.600	2.5
	0.617	1.4		
	0.644	0.0	0.644	0.0
			1.000	0.0

This is a linear relation on a Log–Log plot. To obtain the saturation exponent (n), values of the formation resistivity index (RI) and water saturation (S_w) from laboratory experiments are to be plotted on Log–Log scale. The slope of the fitted straight line (Fig. 9.65) gives the value of n. A standard value of 2.0 may be assumed for the exponent n if laboratory estimated value is not available.

From the above discussion, the basic methodology for estimating water saturation can be established. This methodology makes use of open-hole log analysis and laboratory measurements to simply calculate the water saturation at any point in a reservoir penetrated by some well. The work flow in Fig. 9.66 describes this methodology. It is to be mentioned that the formulae described in this section are for clean

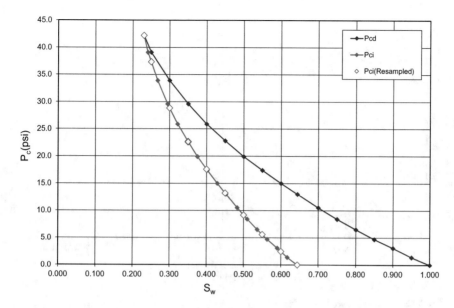

Fig. 9.61 Synthesized imbibition capillary pressure curve

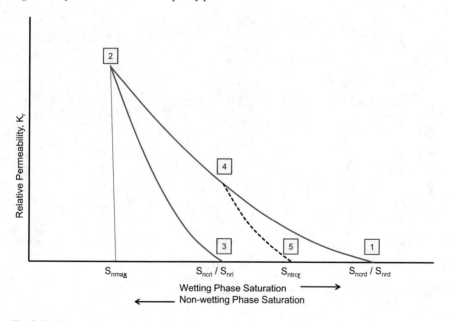

Fig. 9.62 Hysteresis, non-wetting phase

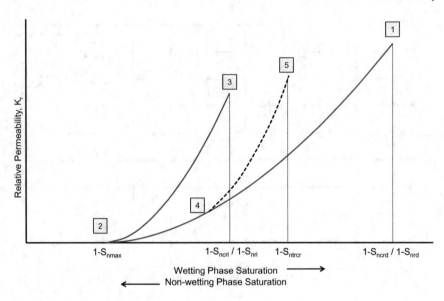

Fig. 9.63 Hysteresis, wetting phase

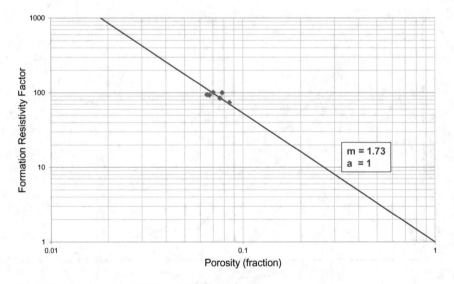

Fig. 9.64 Formation resistivity factor versus porosity

formation. Modified formulae and methodologies are available for handling more complicated lithology. However, describing such formulae and methodologies are beyond the goal of this text.

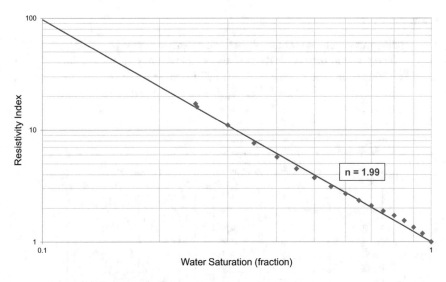

Fig. 9.65   Resistivity index versus water saturation

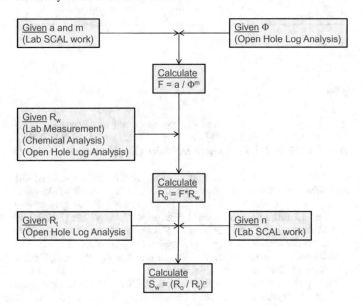

Fig. 9.66   Electrical properties of reservoir rock water saturation evaluation

Exercise

13   A discovery well penetrated oil bearing clean sand zone (4000–4052 ft.ss).
      The given figure shows the resistivity and porosity profiles along that zone.
      The formation water resistivity was estimated as 0.15 ohm.mt. Estimate the
      water saturation at depths 4008, 4012, 4020, 4036 and 4040 ft.ss.
         (n = 2)

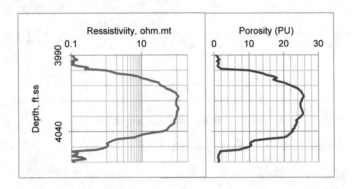

References

Archie, G. E. (1942). The electrical resistivity log as an aid in determining some reservoir
    characteristics. *Petroleum Transactions of AIME, 146*, 54–62. https://doi.org/10.2118/942054-g
Aziz, K., & Settari, T. (1979). *Petroleum reservoir simulation.* Applied Science Publishers, London,
    p. 30.
Baker, L. E. (1988). "Three-phase relative permeability correlations", Enhanced Oil Recovery
    Symposium (Paper SPE/DOE 17369), April 17–20. *SPE, Tulsa, OK, 1988*, 539–554.
Fawakhiri, A. Y., Hassan, I. M., & Abed, A. R., Procedure for initializing reservoir simulation
    model. *Society of Petroleum Engineers*, SPE 17939. https://doi.org/10.2118/17939-MS,1989.
Fayers, F. J., & Mathews, J. D. (1984). Evaluation of normalized Stone's methods for estimating
    three-phase relative permeabilities. *Society of Petroleum Engineers Journal, 24*, 224.
Mandal, D. (2007). *Relative permeability, understanding and usage.* Petronas, Carigali Sdn Bhd
    Reservoir Engineering CoP Presentation, February 2007.
Ramli, A. S., Mustapha, S., Anasir, N., Mohammad, S. Z. (2011). *Formulation of relative perme-
    ability correlation, Malaysian fields, Phase II.* Petronas Carigali Sdn Bhd, Reservoir Engineering
    Studies, April 2011.
Stone, H. L. (1970). Probability model for estimating three-phase relative permeability. *Journal
    Pet. Techology* (Feb.), 214, 1970.

Bibliography

Amott, E. (1959). Observations relating to the wettability of porous rock. *Transactions AIME,216*, 156.

Anderson, W. G. (1986). Wettability literature survey—Part 2 : Wettability measurement. *Journal PET Techology*, 1246,Nov. 1986.

Ewing, J. A. (1989). On hysteresis in the relation of strain to stress. *British Association Reports, 502*, 1989.

A. M. Badawy and T. A. A. O. Ganat, *Rock Properties and Reservoir
Engineering: A Practical View*, Petroleum Engineering,
https://doi.org/10.1007/978-3-030-87462-9

Index

A
Absolute or total porosity. *See* 18
Absolute permeability. *See* 38
Aquifer. *See* 3
Arbitrarily. *See* 131
Arithmetic. *See* 50
Arithmetic-harmonic. *See* 53

B
Basin. *See* 3
Burial depth. *See* 20

C
Capillary attraction. *See* 82
Capillary pressure. *See* 81
Capillary repulsion. *See* 82
Cementing and/or interstitial materials. *See* 20
Combined. *See* 1
Combined rock fluid properties, The. *See* 130
Compaction. *See* 19
Compressibility. *See* 57. *See* 58
Conductivity (C). *See* 180
Connate Water Saturation (S_{wc}). *See* 29
Contact angle. *See* 75
Critical Gas Saturation (S_{gcr}). *See* 30
Critical Water Saturation (S_{wcr}). *See* 30

D
Darcy. *See* 35
De-normalization. *See* 120

Differential traps. *See* 1. *See* 3

E
Effective Formation Rock Compressibility. *See* 58
Effective or interconnected porosity. *See* 18
Effective permeability. *See* 38
Engineering. *See* 52
Enumeration or explicit option. *See* 32

F
Formation density. *See* 24
Formation resistivity factor (F). *See* 182
Fractures and/or vugs increase. *See* 20
Free water level. *See* 91

G
Gas mean free path. *See* 41
Gas slippage. *See* 40
Geometric. *See* 52

H
Harmonic. *See* 52
Harmonic-arithmetic. *See* 53
Hydrostatic equilibrium option. *See* 31
Hysteresis. *See* 131. *See* 177

I
In-parallel. *See* 50
In-series. *See* 52

© The Editor(s) (if applicable) and The Author(s), under exclusive license
to Springer Nature Switzerland AG 2022
A. M. Badawy and T. A. A. O. Ganat, *Rock Properties and Reservoir
Engineering: A Practical View*, Petroleum Engineering,
https://doi.org/10.1007/978-3-030-87462-9

Interfacial tension. *See* 79
Irreducible water saturation (S_{wirr}). *See* 29

K
Klinkenberg slip factor.. *See* 42

L
Leverette (1941) J-Function. *See* 95

M
Must. *See* 130

N
Neutron log. *See* 24
Newton-Raphson. *See* 43
Normalization. *See* 120

O
Oil water contact. *See* 92
Overburden pressure. *See* 58

P
Percussion sidewall. *See* 11
Permeability. *See* 35
Pore space compressibility. *See* 58
Porosity. *See* 17
Primary porosity. *See* 17

R
Relative permeability. *See* 38. *See* 117

Remaining Oil Saturation (ROS). *See* 30
Reservoir. *See* 2
Reservoir pressure. *See* 59
Reservoir simulation modeling. *See* 6
Residual Gas Saturation (S_{gr}). *See* 30
Residual Oil Saturation (S_{or}). *See* 30
Resistivity Index (RI) B. *See* 184
Resistivity (R). *See* 180
Rock bulk compressibility. *See* 58
Rock Matrix (Grains). *See* 58
Rock typing. *See* 127
Rotary sidewall. *See* 11

S
Saturation functions. *See* 130
Scientific modeling. *See* 4
Secondary porosity. *See* 17
Secondary porosity index. *See* 23
Simulation. *See* 4
Sonic log. *See* 23
Sorting. *See* 20
Structural, stratigraphic. *See* 1
Surface tension. *See* 79

T
Three phase relative permeability. *See* 123
Tortuosity. *See* 47
Transition zone. *See* 92
Trap. *See* 1
Type of packing. *See* 20

W
Wettability. *See* 75

Printed in the United States
by Baker & Taylor Publisher Services